Cell Biology:
Review for New National Boards

Cell Biology:
Review for New National Boards

Mark R. Adelman, Ph. D.
Associate Professor of Anatomy and Cell Biology
Uniformed Services University of the Health Sciences
Bethesda, MD

Kurt E. Johnson, Ph. D.
Professor of Anatomy
George Washington University Medical Center
Washington, D.C.

J&S

J&S Publishing Company Inc., Alexandria, Virginia

J&S

Composition and Layout: Ronald C. Bohn, Ph. D.
Cover Design: Kurt E. Johnson, Ph. D.
Printing Supervisor: Robert Perotti, Jr.
Printing: Goodway Graphics, Springfield, Virginia

Library of Congress Catalog Card Number 95-075532

ISBN 0-9632873-8-9

10 9 8 7 6 5 4 3 2 1

Dedication

Mark R. Adelman dedicates his efforts on this book to Abigail and Bennett and time at Beside the Point. Kurt E. Johnson dedicates his portion of this book to John P. Trinkaus, Ph. D., Emeritus Professor of Biology, Yale University. Trink is a remarkable man with enviable energy and wit. Thank you for your guidance, friendship, and parties.

Table of Contents

Preface

This book is designed to enable you to review in just 1-2 days all of the basic cell biology you studied in the first year of medical school: cell membrane phenomena, extracellular matrix, nuclear structure and function, cytoplasmic structure and function, regulation of cell function, and cell biology of diseases. You will find certain topics covered by several sets of questions (e.g. G-proteins). This was done intentionally due to the great importance of the topic to the discipline. We presume that these topics will be covered on examinations because of their importance. Cell Biology is a multidisciplinary body of knowledge, including subjects traditionally covered by other courses. Thus, some of our questions have already appeared previously in Microbiology: Review for New National Boards, Biochemistry: Review for New National Boards, and Physiology: Review for New National Boards. We have been able to condense a review of all basic cell biology into a single book because of the new format of the National Board Part 1 Exam. It is no longer prudent to review exhaustively the basic science courses because the new examination format no longer rewards an encyclopedic knowledge of the basic sciences. Instead, the new exams test knowledge of the scientific basis of disease and the ability to apply basic scientific information to the clinical reasoning process. Consequently, the most efficient way to study for the new exam is 1) to review only the most clinically relevant material from each basic science course and 2) to focus on the application of this material to the solution of clinical problems. These two new study features form the core of this text.

If you answer every question and read all the tutorials in this book, you can cover within 2 days all of the most clinically relevant information from your basic cell biology courses. You will find that many cell biologic facts reviewed or learned anew will be presented in the context of a clinical case or an illustration. We hope that the clinical cases and illustrations will enhance your understanding and recall of the information. Finally, you will learn from the tutorials how cell biological information is used by knowledgeable physicians to understand the courses of diseases and the significance of abnormal findings.

Mark R. Adelman, Ph. D.
Bethesda, MD
Kurt E. Johnson, Ph. D.
Washington, D.C.
March, 1995

Acknowledgements

The authors would like to thank Ronald C. Bohn, Ph. D., Associate Professor, Department of Anatomy, The George Washington University Medical Center for his assistance in formatting the final documents for publication. Gerald V. Stokes, Ph. D., and Ajit Kumar, Ph. D. contributed to materials in the text.

Disclaimers

Figure Credits

Figure 1.1. From J.C. Sherris, ed., Medical Microbiology, An Introduction to Infectious Diseases, ed. 2, ©1990 reprinted by permission of Appleton & Lange, Norwalk, CT and John C. Sherris.

Figure 2.3. From D.W. Fawcett, The Cell, ed. 2, ©1981 reprinted by permission of W.B. Saunders, Philadelphia, PA.

Figure 3.5. The transmission electron micrograph was kindly provided by J. David Robertson, M.D., Ph.D., Department of Neurobiology, Duke University Medical Center.

Figure 4.7. From T.L. Lentz, Cell Fine Structure, ©1971 reprinted by permission of W.B. Saunders, Philadelphia, PA.

Figure 4.9 From K.E. Johnson, Histology: Microscopic Anatomy and Embryology, ©1982 reprinted by permission of Harwal Publishing Company, Malvern, PA.

Figure 6.1. From A.V. Hoffbrand and J.E. Pettit, Essential Haematology, ed. 3, ©1993 reprinted by permission of Blackwell Scientific Publications, London.

Figure 6.2. The transmission electron micrograph was kindly provided by Robert Josephs, Ph. D., Department of Molecular Genetics and Cell Biology, University of Chicago.

CHAPTER I
PLASMA MEMBRANE, CELL SURFACE, CELL ADHESION, AND EXTRACELLULAR MATRIX

Items 1-5

Figure 1.1 is a diagram of a typical bacterial cell with various structures labeled. Match the labeled structure with the most appropriate functional or structural description of the structure in the items below. Answers may be used once, more than once, or not at all.

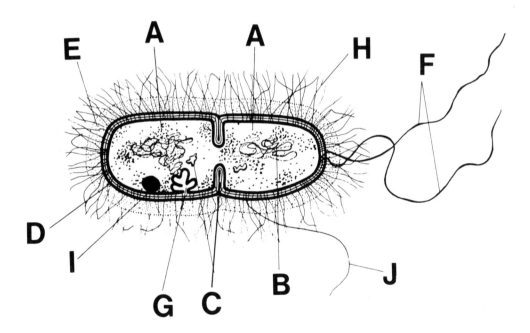

Figure 1.1

1. The genetic information for replication of this organism is found here.

2. Rapid movement of this structure propels the entire organism through its aqueous environment.

3. This organelle prevents macrophages from engulfing the organism.

4.   Food reserves are stored in this organelle.

5.   As this structure extends, the bacterium becomes divided into two daughter cells.

ANSWERS AND TUTORIAL ON ITEMS 1-5

The answers are: **1-B; 2-F; 3-E; 4-I; 5-C. Figure 1.1** shows a typical bacterial cell. **Procaryotes** are functionally compartmentalized, but lack the well-defined membrane bound organelles of eukaryotic cells. Areas of high DNA content appear as areas of high electron density in electron micrographs. This **nucleoid region** (B) contains the bacterial DNA and newly transcribed RNA's. Translation of the information in mRNA into proteins takes place throughout the cell cytoplasm on **ribosomes** (A). After DNA replication and the separation of the duplicated chromosome to opposite ends of the cell, cytoplasmic division occurs as the result of growth of the **septum** (C).

Bacteria move by rotation of **flagella** (F), which may be situated at one end or along the entire length of the bacterium. Smaller appendages called **pili** are found on some bacteria. They are not involved in motility but probably play a role in conjugation or adherence to tissue receptors.

The **bacterial capsule** (E), when present, usually consists of complex peptidoglycans. It serves as a **virulence factor** that prevents phagocytosis by host macrophages or mediates adhesion to host tissue receptors. The antigenic properties of the capsule are sometimes useful in identifying different species and strains.

Many bacteria contain a variety of intracytoplasmic **storage vacuoles** (I). Polymerized metaphosphates or lipids are examples of inclusions which serve to store energy reserves for the organism. The presence of different inclusions or granules can be detected using special stains. Other structures represented within the diagram are a **cell wall** (D), **mesosome** (G), **pili** (H) and **sexual pili** (J).

Items 6-12

The following set of items pertains to cellular membranes and their biochemical composition.

6.   The fluid mosaic model of biological membranes states that

    (A)   All membranes have a core, made up of globular protein subunits, surrounded by multiple layers of hydrophobic phospholipids.

    (B)   Membranes appear to have repeating mosaic patterns because their lipids form micelles that are cross-linked together by a protein backbone.

    (C)   Membrane proteins can move about within the plane of the membrane because the lipids which form the core of the membrane can diffuse relatively freely with respect to adjacent lipid molecules.

    (D)   Transfer of molecules across the membranes usually occurs through channels - lined by phospholipids - that form a repeating, or mosaic, pattern.

    (E)   Proteins can only associate with the lipid core of the membranes by forming covalent linkages with the fatty-acid side chains of constituent phospholipid molecules.

7.   Which of the following is **TRUE** with regard to membrane proteins?

    (A)   Proteins associated with the plasma membranes of cells are highly hydrophobic.

    (B)   Membrane proteins that are in contact with the extracellular matrix are rarely if ever glycosylated.

    (C)   Transmembrane proteins must have lipid molecules covalently coupled to the protein backbone.

    (D)   Membrane channels, through which charged molecules may pass, are frequently formed by self-association of transmembrane proteins.

    (E)   Proteins that are associated with the extracellular face of the plasma membrane can usually only be released from the membrane by treatment with proteolytic enzymes.

8.   Which term is most accurately used to characterize the **interior** of the lipid bilayer of the plasma membrane?

    (A)   hydrophilic
    (B)   acidophilic
    (C)   basophilic
    (D)   amphipathic
    (E)   hydrophobic

9. All of the following are integral membrane proteins of the erythrocyte membrane **EXCEPT**:

   (A) glycophorin
   (B) band 3 protein
   (C) spectrin
   (D) glucose translocase
   (E) Na^+-K^+ ATPase

10. Membrane lipids have all of the following characteristics **EXCEPT**:

    (A) absence of charged groups
    (B) amphipathic
    (C) fatty acids esterified to glycerol
    (D) covalently bound oligosaccharides
    (E) extensive hydrophobic domains

11. All of the following statements concerning membrane composition are true **EXCEPT**:

    (A) Plasma membranes have more lipid than do mitochondrial inner membranes.
    (B) Intracellular membranes have less protein than does myelin.
    (C) Cholesterol is present in plasma membranes and intracellular membranes.
    (D) Cholesterol composition affects membrane fluidity.
    (E) Cell membranes have no free carbohydrate moieties.

12. Which of the following membranes has the **highest** ratio of lipid-to-protein?

    (A) red cell membrane
    (B) myelin
    (C) Golgi membranes
    (D) rough endoplasmic reticulum
    (E) outer mitochondrial membrane

ANSWERS AND TUTORIAL ON ITEMS 6-12

The answers are: **6-C; 7-D; 8-E; 9-C; 10-A; 11-B; 12-B. Cellular membranes** are **lipid-protein bilayers,** consisting of a core bilayer of lipids with which various proteins are associated. The **hydrophobic** tails of the **amphipathic** lipids forming the core of the bilayer associate with one another by hydrophobic interactions. These core lipids are weakly associated with one another within the plane of the membrane; hence, while the bilayer membrane strongly resists separation in a direction perpendicular to the membrane plane, lipid-lipid and lipid-protein associations are relatively labile (fluid) within the plane of the membrane.

Proteins that are specifically associated with membranes may be loosely bound, bound firmly via electrostatic interactions, or so tightly bound that they can only be released by treatment with detergents that completely dissolve the membrane. Proteins that can only be released by detergent treatment are referred to as **integral membrane proteins.** In many cases, such integral membrane proteins extend from one side of the membrane to the other side. Such **transmembrane proteins** are frequently responsible for selective transport of specific hydrophilic molecules across the membrane, thus facilitating passage of such molecules through the hydrophobic lipid bilayer core. The plasma membrane surrounds each cell and is the boundary between the external and the internal environments of the cell.

Deep within cells, there are other (intracellular) membranous systems, including the **rough and smooth endoplasmic reticulum,** the **Golgi apparatus** and the **nuclear envelope.** Individual intracellular organelles such as **lysosomes** are also surrounded by membranes. **Mitochondria** have an **outer membrane** which compartmentalizes the organelle from the cytoplasm. The outer mitochondrial membrane surrounds an **inner mitochondrial membrane.**

The **red cell membrane** is one of the most well characterized plasma membranes. In general, integral membrane proteins span the entire thickness of the plasma membrane and can not be released from their associated lipids without strong detergent treatment or solvent extraction of membranes. They often have a carbohydrate rich portion on the external surface, a middle region rich in hydrophobic amino acids which interact strongly with the hydrophobic domain of membrane lipids and a second hydrophilic domain on the internal (cytoplasmic) surface. The inner hydrophilic domain often interacts with cytoskeletal proteins in the cytoplasmic domain. In the red blood cell, **glycophorin** is an integral membrane glycoprotein of unknown function. **Band 3** is a another integral membrane protein. It interacts strongly with the cytoskeletal protein ankyrin and may be involved in facilitating diffusion of anions across the plasma membrane. **Glucose translocase** is an integral membrane protein that actively transports glucose into red blood cells. **Spectrin** is a cytoplasmic protein that interacts with both glycophorin and band 3 protein; spectrin is thought to have a cytoskeletal function, strengthening the cortex of erythrocytes. **Ankyrin** stabilizes the interaction between spectrin and band 3 protein.

Membrane lipids consist largely of **phosphoglycerides, sphingomyelin, cholesterol** and "other" lipids, e.g., carbohydrate-containing **glycolipids** such as cerebrosides and gangliosides. The phosphoglycerides, sphingomyelin and glycolipids, are extended amphipathic molecules with hydrophilic charged portions containing phosphate and other water soluble moieties such as choline, ethanolamine, serine and inositol esterified to long, hydrophobic, aliphatic hydrocarbon chains. The **amphipathic membrane lipids** form a bilayer where the polar heads of one layer interact with the aqueous external milieu and the polar heads of the other half of the bilayer interact with the aqueous cytoplasmic environment. The hydrophobic portions of each half of the bilayer face one another and interact strongly with one another, with cholesterol and with the hydrophobic portions of integral membrane proteins.

Cholesterol is a compact, hydrophobic, planar compound. It is an important constituent of many membranes even though it is less abundant that phosphoglycerides in the plasma membrane. Cholesterol plays a significant role in determining membrane fluidity. At physiological concentrations, cholesterol limits membrane fluidity. Abnormally high plasma membrane cholesterol levels significantly decrease membrane fluidity. In red blood cells, when the membrane fluidity decreases, these cells become more susceptible to splenic destruction, leading to anemia.

Different cellular membranes have striking differences in their chemical composition. For example, the **myelin sheath** of nerve fibers (a highly modified plasma membrane) is about 80% lipid and 20% protein. The high lipid content of myelin relates to its insulation function in nerve conduction. In contrast, the **inner mitochondrial membrane** is about 80% protein and 20% lipid. Here, there are large numbers of enzymes dedicated to electron transport. Most other plasma membranes are just under 50% lipid and just over 50% protein. Similarly, the lipid composition of different intracellular membranes varies considerably.

Items 13-15

The following set of items pertains to membrane transport phenomena.

13.   All of the following substances move across the plasma membrane by simple diffusion **EXCEPT**:

    (A)   testosterone
    (B)   estrogen
    (C)   aldosterone
    (D) Na^+
    (E)   palmitic acid

14.   All of the following substances move across the plasma membrane by mediated transport **EXCEPT**:

    (A)   testosterone
    (B)   glucose
    (C)   aspartate
    (D)   glutamate
    (E)   galactose

15. Which statement best characterizes the difference between active mediated transport (AMT) and passive mediated transport (PMT)?

    (A)   Only PMT requires energy input.
    (B)   Only AMT can occur against a concentration gradient.
    (C)   Only AMT shows saturation kinetics.
    (D)   Only PMT shows specificity for solute transport.
    (E)   Only AMT can be inhibited.

ANSWERS AND TUTORIAL ON ITEMS 13-15

The answers are: **13-D; 14-A; 15-B**. The plasma membrane is a **selective barrier** between the inside and the outside of individual cells. Solutes can cross the membrane by diffusion or by mediated transport. **Diffusion** is relatively slow for molecules such as glucose because of their low lipid solubility. In contrast, steroids and fatty acids have relatively high lipid solubility and are able to cross the plasma membrane by diffusion. For simple diffusion, there is a linear relationship between rate of transport and the ratio of external concentration of solute/internal concentration of solute, i.e., the higher the ratio, the greater the rate of transport. **Mediated transport** shows saturation kinetics: as solute concentration increases, the rate will reach a maximal velocity when the carrier for mediated transport becomes saturated with solute. Ions such as Na^+, K^+, Ca^{2+} (cations), HPO_4^{2-}, Cl^- and HCO_3^- (anions) as well as amino acids and sugars are transported by mediated transport.

    Mediated transport is facilitated by transporter membrane proteins (translocases) and can be either passive or active. Both passive mediated transport and active mediated transport show saturation kinetics, show specificity for solute transport and can be inhibited. In **passive mediated transport**, solute moves from a high external concentration to a low internal concentration without the expenditure of energy. In **active mediated transport**, solute can move against a concentration gradient because energy (the energy of hydrolysis of ATP) is generated by the Na^+-K^+ ATPase to which the transport is coupled.

Match each numbered biochemical description of collagen with the most appropriate lettered collagen type.

(A)   Type I collagen
(B)   Type II collagen
(C)   Type III collagen
(D)   Type IV collagen
(E)   Type V collagen
(F)   Type VI collagen
(G)   Type VII collagen
(H)   Type VIII collagen
(I)   Type IX collagen
(J)   Type X collagen

16.   This type of collagen is abundant in hyaline cartilage. It contains three identical α chains.

17.   This type of collagen is abundant in bone, the dermis, and tendons. It contains two different α chains.

18.   This type of collagen is the most abundant type in basement membranes.

19.   This type of collagen is rich in carbohydrate. It is abundant on the cell surface of many cell types.

20.   This type of collagen is abundant around blood vessels and smooth muscle in the gastrointestinal tract. It contains three identical α chains.

21.   This type of collagen forms fibrils involved in anchoring basement membranes to underlying connective tissues. It contains three identical α chains.

22.   This type of collagen is abundant in cartilage. It contains three **different** α chains.

ANSWERS AND TUTORIAL ON ITEMS 16-22

The answers are: **16-B; 17-A; 18-D; 19-E; 20-C; 21-G; 22-I. Collagen** is the major extracellular protein of most connective tissues. It is secreted by fibroblasts as well as other cell types and occurs as a large number of molecular variant types, each designated by a specific Roman numeral. Each collagen type consists of three intertwined helical α chains. There are three major different types of α chains (designated α1 to α3) and several minor variants of each of these, based on slight differences in amino acid composition. Different collagen types consist of different combinations of α **chains**.

Type I collagen (A) is widely distributed in the body. It is most abundant in bone, skin, tendons, and dentin. It forms large striated fibrils and is composed of two α1(I) and one α2(I) chains.

Type II collagen (B) is most abundant in hyaline cartilage. It is also present in the vitreous body of the eye and the nucleus pulposus of intervertebral discs. It is composed of three α1(II) chains.

Type III collagen (C) is also widely distributed, like Type I collagen, but is not abundant in bone. It is abundant in fetal skin and surrounding many smooth muscle cells. Type III collagen is also abundant in reticular fibers. It forms small diameter striated fibrils and is composed of three α1(III) chains.

Type IV collagen (D) is abundant in basement membranes in many locations, e.g., the glomerular basement membrane and surrounding the lens. It forms an anastomosing network of fibrils without striations and is composed of two α1(IV) and one α2(IV) chains.

Type V collagen (E) is widely distributed like Type I collagen but does not form striated fibrils. Instead, it forms thin, nonbanded layers around cells. It is prominent in muscle and tendon sheaths. It is composed of two α1(V) and one α2(V) chains.

Type VI collagen (F) is a minor collagen type with wide distribution. It forms fibrils with striations of 100 nm periodicity. It is composed of two α1(VI) and one α2(VI) chains.

Type VII collagen (G) is widely distributed just beneath many epithelial basement membranes. It forms fibrils that help anchor basement membranes to underlying connective tissue structures. It is composed of three α1(VII) chains.

Type VIII collagen (H) is found surrounding endothelial cells of blood vessels and in the cornea. It is composed of three α1(VIII) chains.

Type IX collagen (I), like Type II collagen, is found in cartilage. It is associated with chondroitin sulfate and its fibrillar form is poorly understood. It is composed of one α1(IX), one α2(IX), and one α3(IX) chains.

Type X collagen (J) is restricted to the zone of hypertrophy of developing bones. It is composed of three α1(X) helices.

For each numbered biochemical description of an extracellular matrix molecule or its receptor, choose the most appropriate lettered molecule. Answers may be used once, more than once, or not at all.

      (A)   Collagen
      (B)   Fibronectin
      (C)   Laminin
      (D)   Chondroitin sulfate
      (E)   Heparan sulfate
      (F)   Hyaluronic acid
      (G)   Integrin
      (H)   Elastin
      (I)   Fibrillin

23. This molecule is composed of subunits assembled into a cruciform aggregate. It binds to heparan sulfate, proteoglycan and type IV collagen. It is an essential structural component of many basement membranes.

24. This molecule is a glycosaminoglycan that has a MW = 1,000,000 or more. It forms the backbone of cartilage proteoglycan aggregates.

25. This molecule typically has approximately 1/3 of its amino acids as glycine and is also rich in proline. In some cases, it forms striated fibrillar aggregates.

26. This molecule forms anastomosing molecular chains. Upon hydrolysis, it releases the unusual amino acid desmosine.

27. This molecule is an important cell adhesion protein. It has a MW $\approx$ 460,000 and contains two separate polypeptides chains of MW $\approx$ 230,000 cross-linked by -S-S-bridges.

28. This molecule is a heterodimeric transmembrane glycoprotein. Different forms exist for recognition of fibronectin and laminin.

29. This molecule is a glycosaminoglycan with a repeating disaccharide containing N-acetyl-D-galactosamine and D-glucuronic acid.

30. This is the most abundant protein in the human body.

ANSWERS AND TUTORIAL ON ITEMS 23-30

The answers are: **23-C; 24-F; 25-A; 26-H; 27-B; 28-G; 29-D; 30-A**. **Collagen** (A) is the most abundant protein in the human body. It exists in fibrous, laminar and amorphous forms. Collagen consists of subunit molecules with three helically intertwined α chains of many different types and combinations of chains. Each α chain is a polypeptide with 1/3 of the residues as glycine; it is relatively rich in proline and lysine.

Fibronectin (B) is a MW $\approx$ 460,000 glycoprotein composed of two similar subunits bound together by -S-S- bridges. It exists in two forms. The cellular form is widely distributed in cells of the connective tissues and on cell surfaces. It has several distinct domains that are involved in binding cells to the collagenous extracellular matrix of connective tissues. The plasma form is dissolved in blood and is involved in the clotting mechanism because of its ability to bind to platelets and to the serum protein fibrin.

Laminin (C) is an important component of the basement membrane under many epithelial cells. This molecule, a large glycoprotein, is composed of two different polypeptide chains with MW = 220,000 and 440,000 assembled into a cruciform aggregate. It binds to heparan sulfate, proteoglycan and type IV collagen.

Chondroitin sulfate (D) is a glycosaminoglycan composed of a repeating disaccharide unit with D-glucuronic acid and N-acetyl-D-galactosamine sulfated at the 4 or 6 positions. It is abundant in hyaline and elastic cartilage where it forms a key component of cartilage proteoglycan.

Heparan sulfate (E) is a glycosaminoglycan composed of a repeating disaccharide unit with D-glucuronic acid (or iduronic acid) and N-acetyl-D-glucosamine (or glucosamine). It is sulfated, but less so than the closely related compound heparin. It is found in blood vessel walls, associated with reticular fibers, in the brain and on the surface of many cells.

Hyaluronic acid (F) is a glycosaminoglycan composed of a repeating disaccharide unit with D-glucuronic acid and N-acetyl-D-glucosamine. Unlike other glycosaminoglycans, it can be found free in the extracellular matrix. It is also an important component of cartilage proteoglycan. It is an extended polymer, often with a MW $>$ 1,000,000.

The **integrins** (G) form a complex family of integral membrane proteins. Each distinct integrin contains different types of α and β subunits which are assembled in many different combinations to confer ligand specificity on the integrin. It is a cell surface receptor for many different extracellular matrix glycoproteins including fibronectin and laminin. Integrins are transmembrane proteins involved in signal transduction from the exterior to the interior of many cells.

Elastin (H) is a complex polypeptide found in abundance in elastic fibers. About 30% of its amino acids are glycine and it is also rich in proline and lysine. Several different lysine residues are chemically modified and then covalently bound together to form cross-links which, when hydrolyzed, yield the unusual amino acids desmosine and isodesmosine.

Fibrillin (I) is a glycoprotein with MW = 350,000. It forms a major component of 8-10 nm beaded extracellular fibrils. These are often closely associated with elastic fibers of the basement membrane.

A 17 year-old woman died of a sudden rupture of an aortic aneurysm during a basketball game. She was tall with long arms and legs and hyperextensibility in her digits. Her parents were apparently normal.

31.   The most likely cause of death was from

    (A)   scurvy
    (B)   Marfan's syndrome
    (C)   familial hyperlipidemia
    (D)   lathyrism
    (E)   congenital heart disease

32.   The extracellular matrix molecule defective or deficient in this disease is most often

    (A)   fibronectin
    (B)   type II collagen
    (C)   hyaluronic acid
    (D)   type I collagen
    (E)   fibrillin

33.   The genetic character of the mutation in this particular case was

    (A)   spontaneous new
    (B)   autosomal dominant, full penetrance
    (C)   autosomal recessive
    (D)   sex-linked recessive
    (E)   sex-linked dominant

34.   The defective gene in this case is located on which chromosome

    (A)   11
    (B)   13
    (C)   15
    (D)   21
    (E)   X

ANSWERS AND TUTORIAL ON ITEMS 31-34

The answers are: **31-B; 32-E; 33-A; 34-C**. A single **autosomal dominant gene** defect on **chromosome 15** leads to formation of qualitatively defective (or quantitatively deficient) **fibrillin** and **Marfan's syndrome**, a disease inherited as an autosomal dominant mutation. Since the parents were apparently normal, this patient represented a new spontaneous mutation. Marfan's syndrome patients are tall and thin with long extremities and hyperextensible digits. Aortic aneurysm is a common cause of death. Defects in the extracellular matrix of the media of large blood vessels such as the aorta can lead to aneurysms and fatal rupture, the most common cause of death in Marfan's syndrome.

Items 35-44

For each numbered statement of function described in the items below, choose the most appropriate lettered component of the junctional complex or associated structure. Answers may be used once, more than once, or not at all.

    (A)   Tight junction (zonula occludens)
    (B)   Adhesive junction (zonula adherens)
    (C)   Desmosome (macula adherens)
    (D)   Hemidesmosome
    (E)   Gap junction (nexus)
    (F)   Microfilaments
    (G)   Tonofilaments
    (H)   Microtubules

35.   This structure consists of a hexagonal array of aqueous channels between cells. These pores allow free passage of ions so that adjacent cells are electrically coupled. Small regulatory molecules such as cyclic nucleotides cam also pass through these structures.

36.   This structure is thought to hold cells together. It consists of paired electron-dense plaques on the cytoplasmic faces of apposed membranes. Anchoring intermediate filaments rich in keratin insert into these plaques.

37.   The outer leaflets of the plasma membranes of adjacent cells come into direct contact. It forms a functional seal between a luminal and adluminal compartment.

38.   F-actin is the most abundant and functionally important constituent of these structures.

39. These structures attach epithelial cells to their subjacent basement membranes.

40. The part of the junctional complex most directly responsible for minimizing fluid flow between epithelial cells.

41. Regulates transfer of ions and small molecules from cell to cell in epithelia.

42. Belt-like regions where plasma membranes are in close proximity (but do not touch one another) and cytoplasmic faces of membranes have microfilaments attached.

43. These contain protein assemblies called connexons.

Choose the **ONE** best response.

44. Which of the following statements about components of the junctional complex is **CORRECT**?

   (A) Gap junctions, made of connexons, are found in all junctional complexes.
   (B) Occluding junctions (zonulae occludentes) are most commonly found between epithelial cells and the underlying connective tissue.
   (C) At desmosomes, intermediate filaments attach to a thickening of the cytoplasmic side of the plasma membrane.
   (D) Neither gap junctions nor desmosomes are found connecting cells other than epithelial cells.
   (E) At hemidesmosomes epithelial cells are anchored to actin microfilaments in the underlying basement membrane.

ANSWERS AND TUTORIAL ON ITEMS 35-44

The answers are: **35-E; 36-C; 37-A; 38-F; 39-D; 40-A; 41-E; 42-B; 43-E; 44-C.** Epithelial tissues define boundaries and establish compartments in the human body. For example, the lumen of the small intestine contains a complicated mixture of digestive enzymes capable of digesting the wall of the small intestine. The contents of the lumen of the GI tract are isolated from the sensitive wall of the gut by a membrane specialization known as the junctional complex.

At the most apical portion of the junctional complex there is a **tight junction** where the outer leaflets of the membranes fuse into a belt-like occluding junction called the **zonula occludens** (A). This structure extends around the apex of the columnar epithelial cells and makes a seal between the lumen and the lateral extracellular fluid environment. In freeze-fracture-etch electron microscopy, the zonula occludens sometimes occurs as an anastomosing network of ridges (points of membrane fusion) representing multiple barriers to movement of molecules from the lumen to the lateral extracellular compartment. Below the zonula occludens there is a divergence of the plasma membrane with a clear separation of 10 to 15 nm. This structure is called the **adhesive junction** (zonula adherens) (B). Here there is simple membrane apposition with variable amounts of electron dense material in the intervening 10-15 nm gap. Numerous actin-rich 6 nm **microfilaments** (F) radiate away from the zonula adherens into the cytoplasm of apposed cells. This structure is usually described as an adhesive junction. The **desmosome** (macula adherens) (C) is found below the zonula adherens, as well as at numerous other points of membrane apposition between adjacent cells. At the desmosome, the plasma membranes diverge to 25-30 nm. There is an intermediate dense line running between the cells. On the inner face of each apposed plasma membrane there is a plaque of electron dense material. Long bundles of keratin-rich 10 nm intermediate filaments (called **tonofilaments**) (G) radiate away from the plaque of electron dense material. The macula adherens is thought to be a structure that holds cells together; it is sometimes likened to a "spot weld". **Hemidesmosomes** (D) are often found on the basal surface of epithelial cells where they anchor these cells firmly to the underlying basement membrane.

The **gap junction** (nexus) (E) is also commonly near the junctional complex. In the gap junction, the outer leaflets of the membranes of adjacent cells approach to within 2 nm, but a small, definite gap remains. Gap junctions are composed of hexagonal arrays of barrel-shaped structures with six protein subunits (**connexons**) arranged around an electron lucent central core. This core is an aqueous channel between closely apposed cells, allowing the free passage of ions and other small molecules between epithelial cells.

Items 45-51

The surfaces of B cells and T cells possess functionally important receptors (antigens) which determine their immunological roles. For each numbered description of structure or function, choose the most appropriate lettered antigen. Answers may be used once, more than once, or not at all.

    (A)   T cell receptor
    (B)   CD3 antigens
    (C)   CD4 antigens
    (D)   CD8 antigens
    (E)   LFA-3
    (F) β_2-microglobulin
    (G)   Class I MHC (Major Histocompatibility Complex)
    (H)   Class II MHC
    (I)   Class III MHC
    (J)   CD2 antigen
    (K)   B7
    (L)   CD28 antigen
    (M)   CD40 antigen

45.   Monomeric heavy chain, MW = 44,000; membrane anchored; three cytoplasmic domains; present on virtually all cells; associated with smaller protein; presents foreign antigen to cytotoxic T lymphocytes.

46.   Polypeptide of MW = 12,000; associated non-covalently with both class I and II MHC.

47.   Two distinct polypeptides (α and β), having variable, diverse and constant segments; found complexed with CD3 proteins.

48.   Two distinct polypeptides, MW = 34,000 and 29,000, both membrane anchored; two cytoplasmic domains each, present on B cells, activated T cells and macrophages; restricts helper functions.

49.   An assemblage of at least four distinct membrane-bound polypeptides, some forming heterodimers, all found in conjunction with the T cell receptor.

50.   Two distinct polypeptides, α and β, MW = 40,000 and 50,000; present on helper and cytotoxic T cells; recognizes MHC embedded antigen.

51. Glycoprotein, MW = 50,000, found on mature T cells, natural killer cells and thymocytes; binds to the lymphocyte function antigen-3 (LFA-3).

ANSWERS AND TUTORIAL ON ITEMS 45-51

The answers are: **45-G; 46-F; 47-A; 48-H; 49-B; 50-A; 51-J**. The antigens present on T cells vary in their molecular composition and function. Many possess multiple domains which determine how they will be situated within the cell membrane and interact with other T cell antigens.

The $\alpha\beta$ **T cell receptor** (A) recognizes the processed antigen fragment presented by the antigen presenting cells through their class II MHC. The T cell receptor interacts with cell surface antigens which functionally define cytotoxic, **CD8** (D), or helper, **CD4** (C), functions. **Class II MHC** (H) antigens, present on immunocompetent cells such as B cells, macrophages and activated T cells, function to restrict helper functions. **Class I MHC** (G) antigens, present on all cells, restrict the targeted killing effects of cytotoxic cells. Both MHC classes form noncovalent associations with the β_2-**microglobulin** (F) found on T cell surfaces.

The **CD3 antigens** (B) comprise at least four different membrane bound polypeptides which associate with the T cell receptor in different orientations. The proteins, designated γ, δ, ϵ and ζ, interact individually, as heterodimers and homodimers under different situations. For example, the CD3 complex is thought to be involved in transmembrane signaling. Cell activation via the CD3 complex results in elevation of intracellular Ca^{2+} and phosphorylation of CD3 subunits. The rearrangement of these antigens may be functionally altered in tumor cells rendering them unrecognizable by cytotoxic T cells. The MHC III antigens (I) consist of C2, C4 and B complement components.

Other cell-to-cell interactions must also occur for T cell activation to occur. Interactions such as the binding of **LFA-3** (E) (on the surface of T cells) to the **CD2 antigen** (J) (on the antigen presenting cell) must occur. Blockage of LFA-3 with antibody prevents T cell signaling. The specific ligand for other T cell antigens, such as CD45, has not been identified.

Recent results suggest that antigens found on tumor cells, e.g. the **B7 antigen** (K), present on the surface of certain mouse tumors, act as a costimulator of cytotoxic T cell activity by binding to the **CD28 antigen** (L) present on cytotoxic T cells. The costimulation, with MHC I and T cell receptor comprising the primary recognition interaction, induces multiplication of T cells able to recognize the previously ignored target cells.

The **CD40 antigen** (M), present on B cells, interacts with its corresponding ligand which is present on activated T cells. This allows B cells to switch from expressing IgM to other immunoglobulins. Different cytokines determine which immunoglobulin will eventually be expressed by the activated B cells.

The following set of numbered items relate to solute transport phenomena. For each, choose the most appropriate lettered answer.

52.   Which of the following membrane transport mechanisms requires the expenditure of metabolic energy in the form of hydrolysis of ATP?

    (A)   simple diffusion of steroids
    (B)   simple diffusion of water
    (C)   facilitated diffusion of valine
    (D)   facilitated diffusion of glucose
    (E)   active transport of Ca^{2+}

53.   All of the following substances are transported by active transport **EXCEPT**:

    (A) Na^+
    (B) K^+
    (C)   glucose
    (D) Ca^{2+}
    (E) H^+

54.   The carrier protein for the Na^+-K^+ pump has all of the following characteristics **EXCEPT**:

    (A)   3 Na^+ binding sites on the interior of the cell
    (B)   2 K^+ binding sites on the exterior of the cell
    (C)   a site with ATPase activity closer to the K^+ binding sites than to the Na^+ binding sites
    (D)   two subunits with MW = 55,000 and MW = 100,000
    (E)   found on all cells of the body

55.   The Na^+-K^+ pump has which most crucial function for all cells?

    (A)   maintains a low extracellular Na^+ concentration
    (B)   maintains a high extracellular K^+ concentration
    (C)   ATP hydrolysis
    (D)   maintenance of cell volume
    (E)   maintenance of cell surface charge

56. All of the following statements concerning membrane transport of Ca^{2+} are true **EXCEPT**:

    (A) Different carrier proteins exist in plasma membranes and mitochondrial membranes.
    (B) Different carrier proteins exist in plasma membranes and endoplasmic reticulum.
    (C) The sarcoplasmic reticulum of muscle cells actively sequesters Ca^{2+}.
    (D) Under normal physiological conditions, the intracellular Ca^{2+} concentration is much greater than the extracellular Ca^{2+} concentration.
    (E) The Ca^{2+} pump is an ATPase.

57. All of the following statements concerning ion transport across epithelial sheets are true **EXCEPT**:

    (A) It occurs in the proximal convoluted renal tubules.
    (B) Na^+ is actively pumped into cells apically and actively pumped out of cell basally.
    (C) It is an ATP consuming process.
    (D) Ions are actively pumped out of the lateral and basal surfaces of epithelial cells.
    (E) Apical junctional complexes prevent ions from diffusing from the lateral compartment into the lumen.

ANSWERS AND TUTORIAL ON ITEMS 52-57

The answers are: **52-E; 53-C; 54-C; 55-D; 56-D; 57-B**. There are three basic mechanisms for transport of solutes across semi-permeable cell membranes: **simple diffusion, facilitated diffusion** and **active transport**. Simple diffusion can occur either by solutes passing through membranes, e.g., lipid soluble steroids; or, by solutes passing through aqueous pores in membranes. Facilitated diffusion requires that there be a carrier protein present. Glucose, other sugars and many amino acids cross cell membranes by facilitated diffusion. Active transport involves movement of substances up an electrochemical gradient. Na^+, K^+, H^+, and Ca^{2+} are all moved by active transport. It is mediated by integral membrane transport proteins called "pumps" and requires the hydrolysis of ATP to drive it.

The Na^+-K^+ pump is an important integral membrane carrier protein which spans the plasma membrane. It is found in the cell membranes of all cells. It consists of two different protein subunits. The larger subunit has a MW = 100,000 with 3 Na^+ binding sites and an ATP binding site on its intracellular side and 2 K^+ binding sites on its extracellular side. When Na^+ and ATP bind on the inside, the ATPase becomes activated and the protein becomes phosphorylated. This causes a conformational change in the carrier protein which expels Na^+ on the extracellular side of the membrane. In this conformation, the Na^+-K^+

ATPase binds extracellular K^+. The binding of K^+ dephosphorylates the ATPase, returning it to its original conformation and expelling K^+ at the intracellular side of the membrane. The function of the smaller subunit with a MW = 55,000 is unclear at present.

The most essential role of this ion pump is to maintain the volume of cells. The cytosol contains many negatively charged proteins and low molecular weight solutes. These bind cations. If the ion pumps were not active in expelling more cations than they took into cells, then water would diffuse into cells (via osmosis), eventually causing them to swell to bursting. The Na^+-K^+ pump also establishes gradients of Na^+ and K^+ across the plasma membrane since it expels 3 Na^+ ions for every 2 K^+ taken in. Since these ion gradients cause resting membrane potential, the Na^+-K^+ pump is called electrogenic.

Under normal circumstances, the intracellular Ca^{2+} concentration is extremely low. The extracellular Ca^{2+} concentration is relatively high. This gradient is maintained by a Ca^{2+} pump which is quite similar to the Na^+-K^+ pump in that it has a carrier protein with ion binding sites and an ATPase activity. The carrier proteins in mitochondria and the endoplasmic reticulum are distinct from those in the plasma membrane. In resting muscle cells, the cytosolic concentration of free Ca^{2+} is 10^{-7} M. In contrast, there is a large store of Ca^{2+} in the sarcoplasmic reticulum. When a nerve action potential reaches the motor end plate of a muscle cell, it causes a depolarization of the sarcolemma which in turn causes an increase in the Ca^{2+} permeability of the sarcoplasmic reticulum. This results in a rush of Ca^{2+} ions into the region of the myofibrils, increasing the cytosolic free Ca^{2+} concentration to 10^{-5} M, a crucial event for initiating the sliding of thick and thin filaments relative to one another. These sliding filaments ultimately result in forceful contraction of the entire cell.

Epithelial sheets in many locations in the human body are capable of polarized transport of ions between different compartments. Examples of these transporting epithelia are found in the gastrointestinal tract, kidney tubules, exocrine glands, the ciliary body of the eye and the choroid plexus in the ventricles of the brain. This polarized transport, for example as seen with Na^+ transport out of the proximal convoluted tubules (PCTs) in the kidney, involves active transport of Na^+ out of cells on the lateral and basal surfaces. The apical surface of PCT epithelial cells is freely permeable to Na^+ and water by diffusion. Thus, when Na^+ is actively pumped out of cells at the lateral and basal surfaces, water exits these cells by osmosis at the same locations. Individual epithelial cells are joined apically by an extensive network of tight junctions that are impermeable to Na^+ ions, to prevent leakage of extracellular Na^+ into the lumen.

Items 58-60

Examine the high power transmission electron micrograph below in **Figure 1.2** and then choose the best answer in the items below. The arrow marks the phospholipid bilayer of the plasma membrane and the cytoplasm of the cell is in the bottom of the picture.

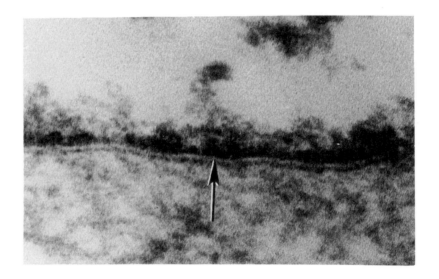

Figure 1.2

58. The electron dense material attached to the outer leaflet of the plasma membrane is best described as the

   (A)   cell cortex
   (B)   contractile ring
   (C)   clathrin-coating
   (D)   glycocalyx
   (E)   nuclear pore complex

59. The most abundant constituent of this layer is

   (A)   actin
   (B)   myosin
   (C)   spectrin
   (D)   glycoprotein
   (E)   phospholipid

60. All of the following characteristics might be attributed to this structure **EXCEPT**:

   (A)   required for cell movement
   (B)   involved in protection of cell from proteolytic digestion
   (C)   involved in immunological recognition
   (D)   involved in cell-extracellular matrix adhesion
   (E)   involved in cell-cell adhesion

ANSWERS AND TUTORIAL ON ITEMS 58-60

The answers are: **58-D; 59-D; 60-A. Figure 1.2** shows the external surface of a cell. Many cells in the human body have a thick coating applied to the outer leaflet of the plasma membrane. This coat is composed of the carbohydrate-rich portions of integral membrane glycoproteins. It is known as the **glycocalyx**. The glycocalyx has several important functions. For example, it is well developed on the apical surface of luminal intestinal epithelial cells. Here, it is thought to prevent the noxious lytic digestive enzymes from autodigesting mucosal epithelial cells. Disaccharidase activity is also part of the intestinal glycocalyx. In other locations, the glycocalyx functions in cell-cell and cell-matrix adhesion. The glycoproteins of the glycocalyx are also involved in immunological recognition phenomena.

Items 61-69

   (A)   Integral membrane proteins
   (B)   Peripheral membrane proteins
   (C)   Both
   (D)   Neither

61.   Associated with membrane lipids by hydrophilic but not hydrophobic interactions.

62.   Often span the entire thickness of the phospholipid bilayer.

63.   Would be numerous in the P-face of a freeze-fracture electron micrograph.

64.   Are often intimately associated with cytoskeletal proteins.

65.   Can have α-helical segments embedded in the hydrophobic domain of the membrane lipid bilayer.

66. Can often be dissociated from membrane preparations by buffers without detergents.

67. Abundant in erythrocytes.

68. Transport proteins in many cell membranes are part of this class.

69. Have little functional significance for the inactive erythrocyte membrane.

ANSWERS AND TUTORIAL ON ITEMS 61-69

The answers are: **61-B; 62-A; 63-A; 64-C; 65-A; 66-B; 67-C; 68-A; 69-D**. Cell membranes are composed of lipids, which are the structural basis for the bilayer of the membrane, and proteins, which are essentially responsible for most membrane functions including ion transport, signal transduction, and cell recognition. Membrane proteins fall into two broad categories: integral membrane proteins and peripheral membrane proteins.

Integral membrane proteins are characteristically rich in hydrophobic domains that interact strongly with the hydrophobic portions of membrane lipids. Thus, they cannot be extracted from membranes easily but require the use of strong detergents that disrupt lipid-protein hydrophobic interactions and solubilize the integral membrane proteins. Some integral membrane proteins do not span the entire thickness of the lipid bilayer but many others are transmembrane proteins that have one or more regions rich in hydrophobic domains, often arranged in α-helices. There may be a single α-helical domain that makes a single pass through the bilayer (e.g., erythrocyte glycophorin A) or there may be many such domains making multiple passes through the membrane (e.g., erythrocyte band 3 which forms HCO_3^- and Cl^- transport channels. In some instances, integral transmembrane proteins are anchored to cytoskeletal proteins. When cell membranes are cleaved in the hydrophobic plane by freeze-fracture, they often leave intramembranous particles exposed in the P-face.

Peripheral membrane proteins can be extracted from membranes by nondisruptive treatments because they are associated with lipids or other membrane proteins by relatively weak ionic interactions. Thus, shifts in the ionic strength or pH of extraction buffers will often solubilize them. Thus, for example, in the erythrocyte membrane, the peripheral protein **spectrin** is associated with another peripheral protein **ankyrin** and a complex between both is ionically associated with hydrophilic groups in the erythrocyte membrane.

The following set of items pertains to membrane fluidity. For each, choose the **ONE** best answer.

70.   Which of the following is the **LEAST LIKELY** event?

   (A)   Polar lipids translocate laterally in the plane of the outer leaflet.
   (B)   Polar lipids rotate about their long axis.
   (C)   Fatty acid side chains of phospholipids vibrate like a tuning fork.
   (D)   Diffusion of integral membrane proteins laterally in the plane of the outer leaflet.
   (E)   Exchange of phospholipids between the inner and outer leaflets.

71.   Which of the following has the most dramatic effect on the rate of diffusion of integral membrane proteins?

   (A)   protein synthesis inhibitors
   (B)   ATP synthesis inhibitors
   (C)   cyanide
   (D)   lowering temperature from 37° C. to 0° C.
   (E)   lowering cholesterol concentration

72.   Which of the following is the **MOST ACCURATE** statement concerning the rate of diffusion of proteins in the plane of membranes?

   (A)   Cholesterol concentrations in membranes have little effect on diffusion rate.
   (B)   Integral membrane protein linkage to the cytoskeleton limits lateral diffusion in living cell membranes.
   (C)   Proteins diffuse in artificial lipid membranes more slowly than in living natural membranes.
   (D)   The rate of diffusion of integral membrane proteins in living biological membranes is approximately 10^{-8} cm^2/sec.

ANSWERS AND TUTORIAL ON ITEMS 70-72

The answers are: **70-E; 71-D; 72-B**. The plasma membrane of eukaryotic cells consists of a bilayer of phospholipids and associated integral membrane proteins and peripheral membrane proteins. The phospholipids are polar molecules with hydrophilic (charged groups in phospholipids) and hydrophobic portions (fatty acids). The charged portions associate with water in either the extracellular face or the cytoplasmic face of the membrane. The hydrophobic portions self-associate in the internal domain of the membrane.

Phospholipids are freely mobile in the plane of the membrane. This occurs rapidly at approximately 10^{-8} cm^2/sec. All other movements described in Item 70 also occur. The energetic requirements to break hydrophilic associations and allow a polar lipid molecule to move between planes of the membrane, however, are the largest. Therefore this movement is least likely. Integral membrane proteins diffuse within membranes without the need for protein synthesis or ATP. Although their mobility would be moderately reduced by decreased cholesterol concentration, clearly, the most dramatic effect would be from lowering temperature.

The rate of diffusion of **integral membrane proteins** in artificial membranes is an order of magnitude (or more) greater than that in living membranes, where diffusion occurs at about 10^{-11} cm^2/sec. The chief factor that limits diffusion of integral membrane protein in living membranes is the anchorage of these proteins to cytoskeletal elements.

Items 73-79

For each numbered description of functional role, choose the lettered cell adhesion protein that **BEST** corresponds. Each lettered answer may be used once, more than once, or not at all.

(A)   N-CAM
(B)   N-cadherin
(C)   Integrin
(D)   TAG-1
(E)   Fasciculin II
(F)   Myelin-associated glycoprotein (MAG)
(G)   Proteolipid protein (PLP)
(H) P_o
(I)   Laminin
(J)   Fibronectin

73.   This cruciform molecule is required for neurite outgrowth. It is also a prominent structural protein in many basement membranes.

74.   This cell adhesion molecule is stabilized by extracellular Ca^{2+} but becomes subject to rapid proteolysis when extracellular Ca^{2+} is removed.

75.   This transmembrane integral membrane protein is a heterodimer of α- and β-subunits. In the $\alpha_5\beta_1$ configuration, it serves as the fibronectin receptor.

76. Intimately associated with actin via vinculin, talin, and α-actinin.

77. This Ca^{2+}-independent cell adhesion molecule is part of the immunoglobulin superfamily.

78. This protein is present on the extracellular face of myelin. It is responsible for the intraperiod line width in myelin of the central nervous system.

79. This myelin protein is highly glycosylated and is responsible for homophilic binding of apposed Schwann cell membranes.

ANSWERS AND TUTORIAL ON ITEMS 73-79

The answers are: **73-I; 74-B; 75-C; 76-C; 77-A; 78-G; 79-H. Neural cell adhesion molecule (N-CAM)** (A) is the most abundant cell adhesion molecule in the nervous system. It is a member of the immunoglobulin superfamily of proteins. N-CAM is used for Ca^{2+}-independent, nonspecific cell-cell adhesion mediated by homophilic interactions, i.e., N-CAMs on adjacent cells interact with one another. **TAG-1** (D), **fasciculin II** (E), **myelin-associated glycoprotein (MAG)** (F), and **L1** are also part of the immunoglobulin superfamily. These N-CAM-related molecules are less widely distributed in the nervous system than N-CAM itself. Their primary role is in directing migration of neurons or glial cells in the developing nervous system by Ca^{2+}-independent, homophilic interactions.

N-cadherin (B) is the most important Ca^{2+}-dependent cell adhesion molecule of the nervous system. N-cadherins on the surfaces of neurons and glial cells mediate cell-cell adhesion by homophilic interactions. Ca^{2+} is thought to induce conformational changes in N-cadherin, thus rendering them less susceptible to proteolytic turnover.

Integrin (C) is another important glycosylated integral membrane protein involved in cell adhesion by a heterophilic mechanism, i.e., interaction between two different constituents, one being the integrin (receptor) and the other being the ligand (fibronectin, laminin, collagen, fibrin, etc.). There are many different classes of integrins based on different combinations of several varieties of α- and β-subunits to form a heterodimer. The ligand specificity of integrins is determined by the specific nature of the heterodimer. For example, the $\alpha_5\beta_1$ heterodimer binds strongly to **fibronectin** (J) in the extracellular matrix. In contrast, the $\alpha_6\beta_1$ heterodimer binds strongly to **laminin** (I), a cruciform glycoprotein of basement membranes. It is currently thought that neurite migration along laminin-rich basement membranes in the developing nervous system is mediated by integrins in neurons. Integrins are connected to actin-rich microfilaments of the cytoskeleton by vinculin, talin, and α-actinin.

Myelin is formed by multiple layers of Schwann cell membranes in the peripheral nervous system (PNS) and of oligodendroglia in the central nervous system (CNS). High resolution transmission electron micrographs of myelin reveal major dense lines (formed by close apposition of the cytoplasmic faces of myelinating cell membranes) and intraperiod lines (formed by apposition of the extracellular faces of myelinating cell membranes). In the PNS, intraperiod lines are 15 nm apart; whereas, in the CNS, intraperiod lines are 14 nm apart. This slight difference is due to differences in cell adhesion proteins.

In the PNS, a transmembrane glycoprotein member of the immunoglobulin superfamily, called P_o (H), binds adjacent layers together by homophilic interactions. The P_o glycoprotein projects slightly further into the extracellular space, thus increasing the dimensions of the intraperiod line. In the CNS, myelin **proteolipid protein** (PLP) (G), a nonglycosylated protein that is unrelated to the immunoglobulin superfamily, projects into the extracellular space slightly less than P_o resulting in a slight decrease in the thickness of the intraperiod line.

Items 80-82

Experimental vascular perfusion of the testis with lanthanum nitrate (an electron dense, low molecular weight tracer) is followed by fixation of seminiferous tubules. Subsequently, seminiferous tubules are prepared for transmission electron microscopy.

80.   Lanthanum nitrate would be found in all of the following locations **EXCEPT**:

   (A)   interstitial tissue
   (B)   surrounding Leydig cells
   (C)   around primordial germ cells
   (D)   around immature primary spermatocytes
   (E)   around spermatids

81.   What structural arrangement prevents penetration of lanthanum nitrate into the adluminal compartment?

   (A)   desmosomes
   (B)   zonula adherens
   (C)   tight junctions
   (D)   gap junctions
   (E)   basement membrane of the seminiferous epithelium

82. These structural elements also function in the formation of all of the following morphological barriers **EXCEPT**:

(A) blood-glomerular filtrate barrier in kidney
(B) blood-brain barrier
(C) blood-bile barrier
(D) impermeable continuous capillaries
(E) barrier preventing leakage of digestive enzymes from intestinal lumen

ANSWERS AND TUTORIAL ON ITEMS 80-82

The answers are: **80-E, 81-C, 82-A**. **Tight junctions** are an essential feature of many epithelial layers. Epithelia line cavities and cover surfaces. They have tight lateral junctions that allow them to serve as boundary tissues, separating one compartment in the body from another. For example, intestinal epithelial cells are joined together by apical junctional complexes. The junctional complex consists of an apical zonula occludens or tight junction, a zonula adherens just deep to the zonula occludens and a macula adherens (desmosome) deep to the zonula adherens. The tight junction is a region of fusion of the outer leaflets of the plasma membranes of adjacent cells. It provides a hydrophobic barrier preventing the contents of the intestinal lumen (digestive enzymes) from diffusing into the lateral spaces between cells. Tight junctions are also present between capillary endothelial cells in continuous capillaries where they serve as the anatomical basis for the blood-brain barrier and between liver parenchymal cells where they serve as the anatomical basis for the blood-bile barrier. The blood-urine barrier in the kidney is more complex. The glomerular basement membrane and the filtration slit diaphragms between podocyte foot processes serve as the anatomical barrier between blood and urine. In the seminiferous epithelium, **Sertoli cells** form a continuous epithelial layer. The spermatogenic cell line is lodged in the spaces between Sertoli cells. A complex web of tight junctions between adjacent Sertoli cells divides the seminiferous epithelium into a **basal compartment** containing only spermatogonia, preleptotene primary spermatocytes and leptotene primary spermatocytes. Later stages in spermatogenesis including late primary spermatocytes, secondary spermatocytes, spermatids and spermatozoa are contained within the **adluminal compartment** apical to the web of tight junctions. This elaborate system of tight junctions prevents exposure of the immune system to foreign antigens of mature gametes. During fetal development, the male gonad becomes equipped with spermatogonia prior to the time in development when the immune system gains the ability to discriminate between native and foreign antigens. Thus, spermatogonia are recognized as native antigens. Spermatogenesis does not begin until puberty, long after the establishment of the immunological sense of native and foreign antigens. Consequently, the surface antigens peculiar to spermatozoa would be recognized as foreign antigens were it not for the tight junctions excluding these foreign antigens from immune surveillance. Other mechanisms also ensure that there is no contact between seminal antigens and the male circulatory system.

The following items deal with the molecular biology of different kinds of membrane proteins. Match the membrane protein in the answers below with the **MOST** appropriate description of its molecular biology in the items below. Answers may be used once, more than once, or not at all.

(A)   Acetylcholine receptor
(B) $GABA_A$ receptor
(C) Na^+-glucose symport protein
(D)   Voltage-gated Na^+ channels
(E) α subunit of Na^+-K^+ ATPase
(F) β subunit of Na^+-K^+ ATPase
(G)   Erythrocyte glucose transporter

83.   Contains on single polypeptide with 11 membrane-spanning α-helical domains.

84.   Contains 5 subunits of 4 varieties.

85.   Single polypeptide which promotes facilitated diffusion of glucose.

86.   Contains an S4 polypeptide rich in basic amino acids.

87.   A single polypeptide with 8 membrane-spanning α-helical domains.

88.   An ion channel that produces inhibitory hyperpolarization of membrane.

89.   Function unknown.

ANSWERS AND TUTORIAL ON ITEMS 83-89

The answers are: **83-C; 84-A; 85-G; 86-D; 87-E; 88-B; 89-F**. The **acetylcholine receptor** (A) consists of five homologous polypeptide subunits with a total MW = 300,000. The 5 subunits occur in 4 varieties, 2 α (with the acetylcholine binding site), 1 β, 1 γ, and 1 δ arranged around a central aqueous pore. When acetylcholine binds to the α subunits, a conformational change occurs which opens the channel briefly, allowing ion flux. All three types of subunits have an α-helical hydrophobic domain which spans the membrane 4 times.

The **GABA_A receptor** (B) also consists of 5 subunits of three types, the probable stoichiometry of which is 2 α, 2 β, and 1 γ. Each subunit has four membrane spanning hydrophobic domains. It is also an ion channel. GABA binding (to α and/or β subunits) causes a conformational change which opens a Cl^- channel, causing inhibitory hyperpolarization of the cell membrane.

The **Na⁺-glucose symport protein** (C) is a single polypeptide with 11 hydrophobic membrane-spanning α-helical domains. Its N terminus is intracellular and its C terminus is extracellular. There is also a large hydrophilic intracellular domain near the C terminus. This membrane protein is responsible for unidirectional coupled transport of Na^+ and glucose.

The **voltage-gated Na⁺ channel** (D) is a single large polypeptide with four large membrane-spanning "subunits", each with 6 hydrophobic α-helical domains. Both the N and C terminals are intracellular. Voltage-gated ion channels have one membrane spanning helix, called S4, which is rich in basic amino acids. Their positive charges interact strongly with the negative charges of membrane fatty acids. In the resting state, the S4 polypeptide spans the membrane. When the membrane becomes depolarized and becomes net positive on the cytoplasmic face, this pushes the S4 helix toward the extracellular side of the membrane, opening the ion channel.

The α **subunit of Na⁺-K⁺ ATPase** (E) consists of 8 membrane-spanning hydrophobic α-helical domains. It has an intracellular Na^+ binding site, and extracellular K^+ binding site, and an intracellular ATP binding site. This membrane transport protein uses the high energy phosphate bond in ATP to drive coupled Na^+ and K^+ transport.

The β **subunit of Na⁺-K⁺ ATPase** (F) is a small glycoprotein with a single hydrophobic α-helical domain. Its function is unknown.

The **erythrocyte glucose transporter** (G) is a large polypeptide with 12 membrane-spanning hydrophobic α-helical domains. The N and C terminals are intracellular. Between the first and second hydrophobic domains, there is an extracellular glycosylation site. Glucose enters erythrocytes by facilitated diffusion, a process that is aided by this glucose permease.

<u>**Items 90-93**</u>

There are common features to many intercellular adhesions. These include involvement of cytoskeletal elements, transmembrane linker glycoproteins, and intracellular attachment proteins. Match the type of intercellular adhesive junction in the answers below with the **ONE** best description in the items below.

(A)   Zonula adherens
(B)   Focal adhesion plaque
(C)   Macula adherens
(D)   Hemidesmosome

90.   The transmembrane linker glycoprotein here is integrin.

91.   The crucial cytoskeletal element here is the microfilament and cadherins are the transmembrane linker glycoproteins.

92.   The crucial cytoskeletal element here is the intermediate filament and the intracellular attachment protein is plakoglobin.

93.   The transmembrane linker glycoprotein is a cruciform molecule.

ANSWERS AND TUTORIAL ON ITEMS 90-93

The answers are: **90-B; 91-A; 92-C; 93-D**. The **zonula adherens** (A) is an apical cell-cell adhesive junction between many epithelial cells. It is supported by microfilaments. Homophilic interactions between **cadherins** join cells together at these adhesive junctions. **Catenins** link **microfilaments** to the cytoplasmic domain of cadherins.

The **focal adhesion plaque** (B) is a cell-matrix adhesion site. Cells adhere to extracellular glycoproteins such as **fibronectin** via $\alpha_5\beta_1$ heterodimeric **integrin**, the fibronectin receptor. **Microfilaments** support this adhesion mechanism. These cytoskeletal elements are connected to integrin via the intracellular attachment proteins **vinculin, talin, and α-actinin**.

The **macula adherens** (desmosome) (C) is a strong intercellular adhesions. It uses **desmocollins** and **desmogleins** as the transmembrane linker glycoproteins, **intermediate (keratin) filaments** as the cytoskeletal support elements, and **plakoglobin** and **desmoplakins** as the intracellular attachment proteins.

The **hemidesmosome** (D) binds epithelial cells to the basement membrane underlying the epithelium. One of the major constituents of this basement membrane is the cruciform extracellular matrix glycoprotein **laminin**. A different kind of integrin heterodimer ($\alpha_6\beta_1$) is the **laminin receptor**. The cytoskeletal elements involved in hemidesmosomes are **intermediate filaments**.

CHAPTER II
NUCLEUS, CELL CYCLE,
MITOSIS AND MEIOSIS

Items 94-97

94.   In a typical eukaryotic cell the nucleus

   (A)   has a diameter of between 50 and 250 μm
   (B)   is tightly connected to the plasma membrane by bundles of microtubules
   (C)   contains chromatin, which is a complex of acidic proteins and glycolipids
   (D)   is surrounded by a double membrane, the outer layer of which has ribosomes bound to it
   (E)   is the principal site of protein synthesis

95.   The membrane of the rough endoplasmic reticulum is continuous with

   (A)   peroxisomal membranes
   (B)   the inner membrane of mitochondria
   (C)   mitochondrial outer membranes
   (D)   the outer membrane of the nuclear envelope
   (E)   the plasma membrane

96.   The primary function of the nucleolus involves

   (A)   transfer of ribosomal RNA into the lumen of the nuclear envelope
   (B)   synthesis of most, but not all, ribosomal proteins
   (C)   packaging of mRNA and transfer RNAs into preribosomal subunits
   (D)   assembly of several kinds of proteins, synthesized in the cytoplasm, with ribosomal RNA
   (E)   pairing of daughter chromatids as the S-phase continues

97. The form of chromatin that is called **euchromatin**

   (A) is the most condensed form of chromatin in the nucleus
   (B) comprises the core of all chromosomes found at the onset of mitosis
   (C) usually is found as a thin layer tightly attached to the inner surface of the inner membrane of the nuclear envelope
   (D) represents portions of the chromatin that are relatively decondensed and are sites where transcription into RNA can occur
   (E) can be distinguished from heterochromatin because euchromatin has very few exons as compared to heterochromatin

ANSWERS AND TUTORIAL ON ITEMS 94-97

The answers are: **94-D; 95-D; 96-D; 97-D**. In most eukaryotic cells the largest organelle is the **nucleus**, with a diameter in the range of 5-15 m. The nucleus is surrounded by the **nuclear envelope**, which consists of two membranes that are continuous with one another at the numerous **nuclear pores**. The outer membrane of the nuclear envelope is continuous with the rough endoplasmic reticulum (RER); and, like the RER, has ribosomes bound to its outer (facing the cytoplasm) surface. The principal material within the nucleus is the **chromatin**, a complex made up of DNA and a variety of nuclear proteins that function to define the packaging of the DNA and determine what portions (and when) are accessible to the processes of DNA duplication, repair, and/or transcription into RNA. When chromatin is condensed and unavailable to such processes it is referred to as **heterochromatin**; the highly condensed chromosomes that take part in mitosis are the most extreme example of heterochromatinization. In contrast, **euchromatin** is the term used to describe chromatin found in the relatively decondensed configuration that allows access of the various proteins required, e.g., for transcription into mRNA. The most obvious "organelle" within the nucleus is the **nucleolus**, consisting of specific portions of chromatin that contain the DNA sequences that code for the major ribosomal RNAs. The nucleolus is the site of rRNA synthesis and processing, as well as the site at which ribosomal proteins - made in the cytoplasm and transported into the nucleus via the nuclear envelope pores - are assembled, along with the rRNAs, into the large and small ribosomal subunits.

Items 98-100

The drug 5-fluorouracil (5-FU) is a potent antimetabolite used for cancer chemotherapy.

98.   Which enzyme is inhibited by this drug?

    (A)   DNA polymerase
    (B)   thymidine kinase
    (C)   reverse transcriptase
    (D)   thymidylate synthetase
    (E)   HGPRT

99.   Which compound would accumulate in 5-FU treated patients?

    (A)   dTMP
    (B)   dUMP
    (C)   hypoxanthine
    (D)   xanthine
    (E)   ATP

100.   5-FU works as a cancer chemotherapy agent by

    (A)   depolymerizing microtubules
    (B)   preventing microfilament assembly
    (C)   arresting mitosis at metaphase
    (D)   blocking DNA synthesis
    (E)   blocking DNA transcription

ANSWERS AND TUTORIAL ON ITEMS 98-100

The answers are: **98-D; 99-B; 100-D**. **5-Fluorouracil (5-FU)** is a **pyrimidine analog**. After activation to the nucleotide level by the salvage pathways, 5-FU is converted into FdUMP which is a potent inhibitor of **thymidylate synthetase**. When thymidylate synthetase is inhibited, DNA synthesis is blocked and cell division is not possible. 5-FU can also be converted to FUTP which will subsequently be incorporated into RNA, perhaps explaining some of the nonspecific cytotoxicity of 5-FU. Thymidine and uridine are both required to rescue 5-FU treated cells.

Items 101-105

For each numbered item of an event that occurs during mitosis, choose the lettered stage of mitosis that most closely corresponds.

(A) Prophase
(B) Metaphase
(C) Anaphase
(D) Telophase
(E) None of the above

101. Nuclear envelope breaks down.

102. Condensation of chromatin into distinct chromosomes.

103. Pairing of homologous chromosomes is completed.

104. Separation of "sister" chromatids begins.

105. Chromosomes decondense and nuclear envelope reforms.

ANSWERS AND TUTORIAL ON ITEMS 101-105

The answers are: **101-A; 102-A; 103-E; 104-C; 105-D**. At the start of **mitosis** each chromosome of a somatic cell consists of two **chromatids**, each of which contains a complete copy of the DNA that defines the chromosome. Mitosis assures the separation of the sister chromatids from one another and their partitioning so that each daughter cell receives one chromatid. Mitosis begins in **prophase**, when chromatin becomes condensed into recognizable chromosomes and the nuclear envelope breaks down. As prophase progresses chromosomes become attached to microtubules of the mitotic spindle and, gradually, are moved about until - at **metaphase** - each chromosome is aligned on the metaphase plate, with sister chromatids "facing" opposite poles of the spindle. During **mitotic** prophase homologous chromosomes move independently of one another; it is only during **meiosis** that homologous chromosomes are paired and then separated from one another. At the onset of **anaphase** the forces that hold sister chromatids together are disrupted and, as anaphase proceeds, sister chromatids separate and are moved to the spindle poles. During **telophase** reconstruction of the nuclear envelope occurs at each spindle pole, creating the two daughter nuclei, within which chromosomes decondense to the chromatin patterns characteristic of **interphase** nuclei.

Figure 2.1 is a karyotype of human cells. Examine this photograph and then answer the items below.

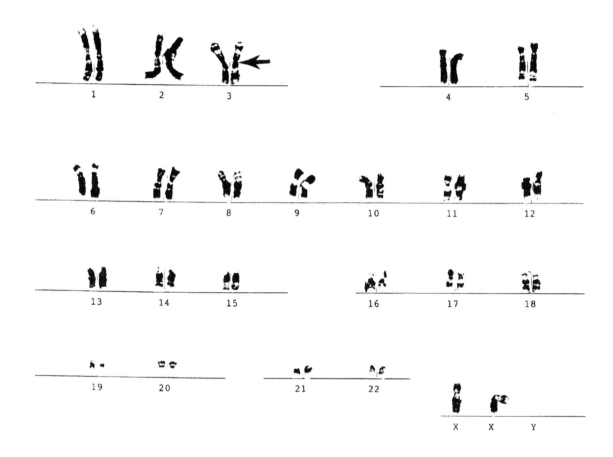

Figure 2.1

106.   Which portion of the cell cycle was this cell in when this karyotype was made?

    (A) G_o
    (B)   S
    (C)   M
    (D)   I
    (E) G_1

107. Which of the following drugs is **MOST COMMONLY** used to treat cultured cells for karyotype analysis?

    (A)   cytochalasin B
    (B)   cytochalasin D
    (C)   actinomycin D
    (D)   colchicine
    (E)   chloramphenicol

108. Which phrase most accurately describes this karyotype?

    (A)   triploid
    (B)   aneuploid
    (C)   trisomic
    (D)   normal, male
    (E)   normal, female

109. The constricted structure at the arrow in chromosome 3 is called a

    (A)   centromere
    (B)   microtubule
    (C)   nucleosome
    (D)   nuclear pore
    (E)   centriole

ANSWERS AND TUTORIAL ON ITEMS 106-109

The answers are: **106-C; 107-D; 108-E; 109-A**. To produce a **karyotype** such as the one shown in **Figure 2.1**, tissue samples, e.g., from punch biopsy, chorionic villus biopsy, or amniocentesis, are cultured *in vitro*, and treated with **colchicine**, a drug that prevents assembly of microtubules and therefore prevents formation of the mitotic spindle. As a result, cells are arrested at **metaphase** (M of the cell cycle). Next, cultures are stained with special formulations of basophilic dyes and squashed to reveal banding patterns in condensed metaphase chromosomes. Finally, the chromosomes are photographed; cut outs from the photographs are arranged in homologous pairs to constitute a karyotype. This particular karyotype has 22 pairs of **autosomes** and a homomorphic pair of X **sex chromosomes**. This is a **euploid** karyotype of a normal female. If it were **aneuploid**, e.g., trisomic for chromosome 21, it would have 3 copies of chromosome 21 and be called Trisomy 21 (Down syndrome). This kind of trisomy is most often caused by meiotic nondisjunction.

The constriction at the arrow is the **centromere**. On either side of the centromere there is a protein-rich structure called the **kinetochore**. Each metaphase chromosome has two kinetochores, one for each chromatid. Each kinetochore is a flat plate with one side attached to the chromatin of each chromatid and on the other side attached to **microtubules** of the **mitotic spindle**. The kinetochore thus is a microtubule attachment site on the chromosome. The microtubules are responsible for moving chromosomes from the metaphase plate toward the centrosomes at opposite poles of a dividing cell. Chromosome movement occurs during anaphase.

Items 110-115

The following set of items pertains to the structure and function of the mitotic spindle. For each numbered statement of function, choose the lettered structure that most closely corresponds.

(A)   Polar microtubules
(B)   Kinetochore microtubules
(C)   Astral microtubules
(D)   Centriole
(E)   Centrosome
(F)   Centromere
(G)   Kinetochore
(H)   Microfilaments
(I)   Kinesin

110.   These tubulin-rich structures extend from the centrosomes to the chromosomes.

111.   These structures are parts of the chromosomes that serve as attachment sites for microtubules.

112.   This structure is a cylindrical organelle that contains 9 triplet microtubule clusters.

113.   These structures are **not** part of the mitotic spindle, but they function - at the end of mitosis - to constrict the cytoplasm so that the daughter cells separate from one another.

114.   An ATPase, this enzyme is involved in moving chromosomes along microtubules.

115.   Anchored to the centrosomes, these structures radiate out into the cytoplasm surrounding the mitotic spindle.

ANSWERS AND TUTORIAL ON ITEMS 110-115

The answers are: **110-A; 111-G; 112-D; 113-H; 114-I; 115-C.** At the onset of mitosis, nuclear chromatin has been condensed into chromosomes, consisting of paired chromatids; each chromosome has a constriction, termed the **centromere**, at which each chromatid has a structure - the **kinetochore** - to which **kinetochore microtubules** attach. With the breakdown of the nuclear envelope at prophase, the mitotic spindle forms. The two poles of the spindle are defined by the presence of a **centrosome**, each of which contains a **centriole** surrounded by amorphous material. Centrioles are cylindrical structures (about 0.2 μm in length) containing nine triplets of microtubules arranged in a manner similar to the arrangement of the outer microtubule doublets of ciliary and flagellar axonemes. Material around the centriole somehow serves to nucleate the assembly of many of the microtubules that make up the mitotic spindle (**polar microtubules**) and radiate away from the spindle (**astral microtubules**) to define its location within the dividing cell. Prominent microtubules that extend from the spindle pole to the immediate vicinity of the chromosomes are called polar microtubules. Many of the microtubules that make up the spindle assemble on the kinetochores (two on each chromosome); these **kinetochore microtubules** orient the chromosomes at the metaphase plate and - normally - guarantee that one of the two chromatids from each chromosome will be carried to each pole, thus assuring that each daughter nucleus will receive one chromatid from each chromosome. The anaphase movement of the separating chromatids toward the spindle poles is powered by motor proteins, as least some of which are **kinesins** - ATPases that direct a number of movements along microtubule tracks. Once **karyokinesis** is complete, **cytokinesis** - the constriction of the cytoplasm in a plane perpendicular to the spindle axis - occurs; thus creating two daughter cells, each of which contains a newly-formed nucleus. The contractile ring that causes cytokinesis is made up of actin-**microfilaments** that - in interaction with a cytoplasmic ATPase of the myosin type - squeeze the cytoplasm transiently connecting the daughter cells in a "purse-string" fashion.

The following set of items pertains to the cell cycle. Match the portion of the cell cycle in the answers with the most appropriate description of this portion of the cell cycle in the items.

(A) G_o
(B) G_1
(C) G_2
(D)   S
(E)   M

116.   Many differentiated cells are permanently arrested in this phase of the cell cycle.

117.   This is the stage of the cell division cycle blocked by cancer chemotherapy drugs designed to block DNA replication.

118.   Colchicine arrests cell in this phase of the cell division cycle.

119.   During this phase, chromosomes become condensed and aligned prior to equal partition into daughter cells.

120.   Oncogenic factors from tumor viruses target the growth suppressor function of this phase of the cell cycle.

121.   Mutagenic events leading to DNA damage are repaired during this phase of the cell cycle.

122.   This is the stage of the cell cycle that immediately precedes mitosis.

ANSWERS AND TUTORIAL ON ITEMS 116-122

The answers are: **116-A; 117-D; 118-E; 119-E; 120-B; 121-C; 122-C**. The growth and division cycle of all somatic cells is divided into four principal phases: **M = mitosis, G_1 = growth phase 1** following mitosis; **S = DNA synthesis phase**; and **G_2 = growth phase 2** immediately following DNA synthesis. The proportion of time spent in each growth phase is dependent upon the physiological state of the cell. Many differentiated cells appear to be in an indefinite arrested state corresponding to an extended G_1, sometimes called the **G_o phase** (A).

To commit a normal cell to re-enter the cell division cycle, the resting G_o cell must receive specific growth stimuli or be induced by oncogenic factors. Once the stimulated cell progresses past a hypothetical "restriction point", it is committed to enter the DNA replication or S-phase (D). Completeness of DNA replication is monitored during the G_2 phase (C). At the end of G_2, cells are committed to enter the M-phase (E) or mitosis; similarly, when cells enter prophase of meiosis I, they do so at the end of G_2. The enzymes and other gene products that monitor and regulate specific phases of the cell growth cycle are of intense interest in seeking more specific and effective targets for cancer therapy, since it is likely that most oncogenic factors derange cell growth by altering the normal functions of the check points within the cell cycle.

Progression of cells through the discrete phases of the growth cycle is controlled at several different molecular check points. Thus, DNA is not replicated throughout the entire cycle, but only during the S-phase. Since cancer cells divide more rapidly than normal cells, they are more often in S-phase and are therefore more susceptible to nucleotide analogs which are competitive inhibitors of DNA replication.

In normal cells, the fidelity of DNA replication is monitored at the S-phase. For example, if the cell is exposed to ionizing radiation with subsequent damage to newly replicated DNA, the cell delays the next phase of the cell cycle, "the G_2 delay", in order to allow time for DNA repair. Without this molecular check point, mutated DNA would be passed on to the next round of DNA replication. Cancer cells lack this G_2 check point.

A class of protein **serine/threonine kinases** are responsible for the regulation of cell cycle events. The catalytic subunit of the kinase, in association with regulatory proteins called **cyclins,** forms the crucial cell cycle regulatory molecules. The cyclins are named because of their proteolytic degradation at the end of each mitotic phase. There are cyclins specific for G_1-, S- and M-phase. Oncogenic proteins induce cell growth by targeting G_1 cyclins. Functional regulation of cell cycle kinases, therefore, forms a crucial target point for control of cell division.

For each numbered item of an event during meiosis, choose the lettered phase that is the most appropriate match for the time during which the event is most likely to occur.

(A)	Prophase I
(B)	Metaphase I
(C)	Anaphase I
(D)	Telophase I
(E)	Cytokinesis after Meiosis I
(F)	Prophase II
(G)	Metaphase II
(H)	Anaphase II
(I)	Telophase II
(J)	Cytokinesis after Meiosis II
(K)	None of the above

123.   Recombination occurs between homologous chromosomes.

124.   Sister chromatids separate.

125.   Homologous chromosomes separate.

126.   Homologous chromosomes are paired and the kinetochores of sister chromatids are "pointed" toward the **same** spindle pole.

127.   Synaptonemal complexes form at this stage.

ANSWERS AND TUTORIAL ON ITEMS 123-127

The answers are: **123-A; 124-H; 125-C; 126-B; 127-A.** Each somatic cell in the human body is **diploid,** having two copies of each chromosome - one maternal and one paternal. Sexual reproduction requires a mechanism for the production of **haploid** cells (gametes - ova and spermatozoa) that, upon fusion at fertilization, yield a diploid cell with genetic material from both parents; **meiosis** is the process that accomplishes this. Prior to meiosis I, each chromosome has been duplicated and consists of two chromatids, joined at the centromere - just as prior to mitosis. However, several events distinguish meiosis I from mitosis. In prophase of meiosis I, homologous chromosomes join together and form - at one or more points - **synaptonemal complexes**, at which crossing over and recombination occurs; this produces chromosomes with mixtures of parental and maternal alleles. At metaphase of meiosis I, homologous chromosomes remain paired, so that the metaphase plate has pairs of maternal and paternal chromosomes (tetrads) oriented such that - at anaphase I - maternal and paternal chromosomes are carried by the meiotic spindle to opposite poles. This guarantees that each daughter cell from meiosis I is haploid (has **either** a maternal copy or a paternal copy of each chromosome) and that each chromosome has two chromatids. Immediately after meiosis I (and the cytokinetic event that separates the daughter cells) cells proceed - without another S phase (i.e. without additional DNA synthesis) into meiosis II. Meiosis II is very much like a "conventional" mitosis except that each cell has only a haploid number of chromosomes. At the metaphase II plate, chromosomes are aligned so that sister chromatids "point" towards opposite poles. Anaphase II then separates sister chromatids and the ensuing telophase II and cytokinesis II yield two daughter cells, each of which is haploid with just one copy (one chromatid) of each chromosome.

Items 128-130

128.   Restriction endonuclease digestion of DNA can be used in all of the following medical situations **EXCEPT**:

    (A)   diagnosis of sickle cell anemia
    (B)   diagnosis of β-thalassemia
    (C)   diagnosis of cystic fibrosis
    (D)   diagnosis of pregnancy
    (E)   establishment of paternity

129.   Restriction fragment length polymorphisms (RFLPs) are medically useful in a growing number of arenas. Which statement **BEST** characterizes these RFLPs?

    (A)   Different restriction enzymes cut DNA at different palindromes.
    (B)   Differences in DNA sequence result in alterations in DNA sensitivity to individual restriction endonucleases.
    (C)   Restriction enzymes degrade different DNA samples into nucleotides at different rates.
    (D)   Loss of a restriction site by gene mutation would increase the number of restriction fragments.
    (E)   They are detected by Western blots.

130. Which statement best characterizes molecular analysis of sickle cell anemia?

    (A)   It is caused by an autosomal dominant mutation.

    (B)   The mutation alters the structure of the α-globin portion of hemoglobin.

    (C)   Endonucleases can not be used in its diagnosis.

    (D)   A codon change destroys a restriction site.

    (E)   It is caused by deletion of the β-globin exon.

ANSWERS AND TUTORIAL ON ITEMS 128-130

The answers are: **128-D; 129-B; 130-D**. **Sickle cell anemia** results from a mutation that changes a glutamic acid residue (codon GAG) to a valine residue (codon GTG) at position 6 in the β-globin chain of hemoglobin. The mutation can be detected by comparing DNA from sickle patients and normal patients after **restriction endonuclease** MstII digestion. MstII cuts at the sequence CCTNAGG (where N is any nucleotide). The A → T change in sickle cell DNA destroys an MstII cutting site, one that normally yields a 1.1 kb fragment of DNA. Wild type MstII digested DNA when probed with a β-globin gene probe yields one 1.1 kb and one 1.3 kb DNA. Sickle cell DNA yields only the 1.3 kb fragment, a difference that is readily detected by the **Southern blot technique**. With this technique, DNA fragments are separated (according to their size) by gel electrophoresis and then transferred to nitrocellulose membranes. Next, the nitrocellulose filter is incubated with radioactively labelled DNA (the probe) that will hybridize to its complementary sequence on the blot. Detection of the radioactive band based on exposure of X-ray films indicates the size of the DNA fragment complementary to the DNA probe. These techniques are precise and sensitive, allowing detection of the sickle trait prenatally. Unfortunately, mutations that both cause a disease **and** alter a restriction endonuclease cutting site are uncommon.

    Genetic linkage, i.e., coinheritance of a disease gene with a detectable marker such as a restriction fragment length polymorphism (RFLP) has been utilized to isolate genes of some important human diseases. For example, the genes for cystic fibrosis, Duchenne's muscular dystrophy and Huntington's disease have all been identified.

    Single base changes of the type described for sickle cell anemia are unfortunately quite common. By some estimates, the frequency may be as high as 1 in 100 base pairs. Most of the single nucleotide changes do not, however, lead to changes in the protein product but they can alter a restriction enzyme cutting site. The result of such an **altered restriction site** will be a "**polymorphism**" in the length of DNA fragments generated by a given restriction enzyme since the individual's normal chromosomal DNA will be cut by the enzyme, whereas the homologous chromosome with the mutated restriction site will not be cut. **Restriction fragment length polymorphisms** (RFLPs) are inherited just like a gene. Consequently, by following the inheritance pattern of a disease in a family and the linkage (coinheritance) of a RFLP, one can identify the gene locus responsible for the disease.

Duchenne's muscular dystrophy (DMD) is an X-linked disorder causing progressive muscular degeneration in young males. DMD occurs in about 1 in 3500 live births and there is no cure. Patients usually die in their early twenties. Using RFLP linkage and the chromosomal translocation of the DMD gene which placed it at chromosome Xp21, the DMD gene has been cloned. It is spread over 2.5×10^6 kb and has at least 65 exons. **Genomic deletion** has been detected in 60% of the patients with DMD and use of genetic probes has been useful in diagnosing DMD.

Cystic fibrosis (CF) is a common (1 in 2500 live births among Caucasians), severe genetic disease characterized by respiratory infections, pancreatic insufficiency and an elevation of Cl⁻ in sweat. Identification of families with the disorder and analysis of DNA from these families with a panel of probes from all parts of the human genome based on RFLP linkage was successful in identifying the CF gene on chromosome 7, and isolation of the DNA sequence which encodes a protein called **cystic fibrosis transmembrane conductance regulator** (CFTR). Recent laboratory experiments have successfully transferred the CFTR gene into CF cells *in vitro* and have lead to correction of the CF cell defects. Identification and isolation of the gene responsible for CF is the first step toward gene therapy for CF, an approach that has already met with some success in the laboratory and which holds great promise for future clinical applications.

Of the some 4,000 known human genetic disorders, in most cases the normal function of the affected gene has not been identified. Linkage analysis of the type described above will play an important role in identifying and cloning the responsible genes.

Items 131 and 132

DNA-based diagnosis and treatment of inherited diseases is rapidly becoming an invaluable tool in the practice of medicine. The following set of items concerns recent advances in molecular biology for the diagnosis and treatment of inherited diseases.

131. All of the following statements concerning β-thalassemia are correct **EXCEPT**:

(A) There are more than 50 different mutations causing β-thalassemia.
(B) β⁰-thalassemias are characterized by a complete lack of β-globin.
(C) β⁺-thalassemias are characterized by overproduction of β-globin.
(D) β⁰-thalassemias can be caused by single base changes leading to translation termination.
(E) β⁺-thalassemias can be caused by mutations in the β-globin promoter region.

132. Which of the following statements is the most accurate description of the molecular basis of hemophilia?

(A) Genetic engineering eliminates the risk of HIV-infection from purified clotting factors.

(B) The gene for factor VIII has been cloned.

(C) Cloned factor VIII genes have been used to synthesize large quantities of factor VIII for treatment of hemophilia.

(D) Hemophilia is caused by factor VIII deficiency which results from an insertion of retrotransposons into factor VIII gene introns.

ANSWERS AND TUTORIAL ON ITEMS 131 AND 132

The answers are: **131-C; 132-D**. **Thalassemias** (α or β) are diseases caused by abnormal synthesis of the α- or β-chains of hemoglobin. Genetic mutations that induce thalassemias represent a spectrum of abnormalities in globin genes including transcription, RNA splicing, mRNA stability and translation. Over 50 different mutations have been reported in the β-globin gene. βo-**thalassemia** is characterized by a complete absence of β-globin caused by either single base mutations that give rise to translation termination or short deletions and insertions in the β-globin coding sequence altering the reading frame. β+-**thalassemia** is characterized by reduced levels of synthesis of β-globin chains. Some β+-thalassemias are due to mutations within the β-globin promoter region which result in an 80% reduction in the levels of β-globin mRNA.

Other β-thalassemias are caused by mutations that lead to abnormal splicing of the globin mRNA. Mutations involving single base substitutions in the critical 5′ or 3′ splice junctions lead to cryptic splice sites resulting in incorrectly spliced mRNA that is either untranslatable or is translated into non-functional proteins. Aberrantly spliced globin mRNA can be detected by S1-nuclease protection assays or Northern blot analysis of the mutant mRNA.

Factor VIII gene mutations, leading to **hemophilia**, are caused by insertions of L_1 nonviral **retrotransposons** in exon 14 of the hemophilia gene, disrupting normal gene function. Hemophilia is an X-linked inherited disorder in the blood clotting mechanism due to defective factor VIII. It is characterized by prolonged bleeding. Until recently, the missing factor VIII was supplied to patients after purification from donated blood. There is a finite risk of inadvertent hepatitis or HIV infections from contaminated donor blood. With the cloning of human factor VIII gene, it has become possible to produce large quantities of factor VIII for hemophilia therapy without the attendant risk of infectious disease from contaminated blood. The cloned gene is also useful as a probe for diagnosis of factor VIII hemophilia.

Defective factor VIII gene arises from the insertion of an L_1 repeat element (a nonviral retrotransposon) into exon 14 of the factor VIII gene. Exon 14 is bordered by two Taq1 restriction sites. The L_1 insert in hemophiliacs introduces an additional Taq1 site within the L_1

element. Digestion of normal DNA from the mother and grandmother with Taq1 restriction endonuclease and comparing a Taq1 digest of hemophiliac DNA by Southern blot using exon 14 cDNA as a probe reveals the additional Taq1 DNA fragment unique to the hemophiliac. This diagnostic test is useful and precise.

Items 133-136

Polymerase chain reaction (PCR) technology has revolutionized DNA-based diagnostic procedures. This technique was developed in the mid-1980s and exploits certain basic features of the DNA replication process.

133. The polymerase chain reaction (PCR) requires all of the following **EXCEPT**:

 (A) highly purified and cloned DNA
 (B) a thermal cycling device
 (C) a supply of deoxyribonucleotide triphosphates
 (D) a heat stable DNA polymerase
 (E) synthetic oligonucleotide primers

134. The chief advantage of the PCR technique is that it

 (A) allows synthesis of mRNA
 (B) allows amplification of specific DNA sequences
 (C) allows insertion of genes that can be replicated
 (D) can be performed at body temperature
 (E) is inexpensive

135. During the high temperature phase of the PCR technique

 (A) mRNA is denatured
 (B) proteins are denatured
 (C) annealing of mRNA and DNA is prevented
 (D) DNA duplexes melt to produce single-stranded DNA template
 (E) the primer oligonucleotides are inactivated

136. PCR can be used in all of the following instances **EXCEPT**:

(A) detection of RAS oncogene mutations
(B) classification of tumors
(C) design of therapeutic chemotherapy protocols
(D) isolating chromosomes
(E) studying the pattern of gene expression

ANSWERS AND TUTORIAL ON ITEMS 133-136

The answers are: **133-A; 134-B; 135-D; 136-D**. The **polymerase chain reaction** (PCR) is a **DNA amplification process** based on synthetic oligonucleotide primers that allow the synthesis of large quantities of a specific stretch of DNA. DNA polymerase requires single-stranded DNA as a template for synthesis of the new complementary strand. Another important feature of DNA synthesis is the requirement of a primer sequence that provides the 3'-OH end for extending the newly synthesized DNA strand.

A unique feature of the PCR is that by supplying synthetic oligonucleotide primers which are designed to anneal to a specific sequence on the template DNA, one can direct the copying of a desired region of the DNA template. The PCR primers are chosen to flank the region of DNA that is being amplified so that new primer sites are generated on each newly synthesized DNA strand. "Amplification" of the DNA flanked by the primer binding site is accomplished by alternate cycles of heating to 94° C to melt off DNA duplex into single stranded DNA and cooling to reanneal the primer oligonucleotides to start a new round of synthesis. The PCR procedure uses a thermocycler, the heat-stable *Taq* (from *Thermus aquaticus*, a microorganism that lives in hot springs) DNA polymerase and the necessary deoxyribonucleotide triphosphates. PCR does not require purification of the DNA template, and can amplify a specific DNA sequence several hundred million fold within a short time. The practical importance of this procedure for diagnostic purposes is only limited by our awareness of the genetic lesions.

PCR can also be used to study the pattern of gene expression by a process called "reverse PCR". Here, mRNA is converted into cDNA (complementary DNA), using reverse transcriptase (RNA-dependent DNA polymerase). The cDNA then serves as the template for the usual PCR procedure. DNA is a relatively stable molecule and the PCR requires only a small amount of unpurified DNA. Therefore, amplification can be performed on old tissue biopsies, even those embedded in paraffin.

A growing number of diseases are now being diagnosed effectively using PCR. For instance, rapid screening of a large population of patients by PCR revealed that different forms of lymphoid tumors had different RAS oncogene mutations. In addition, follicular lymphomas, retinoblastoma, neuroblastoma, lung cancer and breast cancer can be accurately classified using PCR. Accurate diagnosis of certain tumors can help in the design of more effective treatment protocols.

Items 137-144

Molecularly cloned DNA sequences are rapidly becoming an important tool in medicine. They can be used in the production of clinically important proteins, as diagnostic tools or in the application of gene therapy of inherited genetic defects. The following list of terms refer to some key aspect of molecular cloning technology. Match the appropriate terms with their correct definition in the items.

(A)   Genomic library
(B)   Plasmid vector
(C)   Complementary DNA (cDNA)
(D)   Nucleic acid probes
(E)   Expression vectors
(F)   Gene screening
(G)   Transfection
(H)   Reporter gene

137.   Messenger RNA (mRNA) sequences in the cell can be utilized as templates and copied into complementary DNA sequences using a retroviral enzyme, reverse transcriptase.

138.   Independently replicating circular DNA molecules in bacteria that usually carry genes for antibiotic resistance, are used as carriers of cloned genes for amplification in a bacterial host.

139.   Specific nucleic acid sequences, DNA or RNA, made radioactive by enzymatic synthesis using labeled nucleotides, can be utilized to search for complementary DNA/RNA sequences from the entire complement of cellular DNA or RNA.

140.   A collection of recombinant vectors, usually derived from bacteriophage lambda, that encompasses DNA inserts which represent the entire complement of genomic DNA of the cell.

141.   Vectors for high level expression of protein are stitched together with DNA molecules representing essential components of the functional gene under the control of a potent promoter element.

142.   A procedure of transfer of purified genetic material into mammalian cells that involves intake of calcium phosphate precipitate of DNA molecules by recipient cells in culture.

143. In this process, specific nucleic acid probes are used to search for the DNA sequences in a genomic or cDNA library.

144. A specific gene whose product can be assayed easily and whose levels of expression under the control of particular promoter elements is taken as an indication of the efficiency or strength of the promoter/enhancer sequences.

ANSWERS AND TUTORIAL ON ITEMS 137-144

The answers are: **137-C; 138-B; 139-D; 140-A; 141-E; 142-G; 143-F; 144-H. Gene cloning techniques** allow the isolation, modification and insertion of desired genes into another genome. Such recombinant DNA technology has revolutionized the study of biology and has provided medicine with efficient means of producing valuable therapeutic reagents (e.g., Factor VIII for hemophilia).

Restriction endonucleases are useful in gene cloning because they cut double-stranded DNA at precisely defined nucleotide sequences. DNA from a circular bacterial plasmid (typically an episome carrying genes for antibiotic resistance), and chromosomal DNA can be cut by a particular restriction endonuclease, leaving linear plasmid DNA molecules with cohesive ends and a vast array of different chromosomal DNA fragments with complementary free ends. Annealing and DNA ligase treatment then allows formation of plasmid DNA molecules with many different chromosomal DNA fragments inserted and will produce a **genomic library** (A). The genomic DNA library will contain many DNA fragments, only a few of which are complete genes. This becomes a severe problem in eukaryotic genomes, where much of the DNA consists of noncoding sequences (introns). Similar techniques can also be used to introduce chromosomal DNA into viral vectors. The plasmid or virus carrying the added DNA is called a **cloning vector**. When the viral cloning vector is transfected into host cells, it can be replicated to more than 10^{12} copies in a single day. **Transfection** (G) is an inefficient process. Use of plasmids carrying the genes for antibiotic resistance allows selection of transfected cells by growth in media containing antibiotics.

Complementary DNA (cDNA) libraries (C) are much more useful for eukaryotic cloning strategies. In this technique, the synthesis of complementary DNA sequences is accomplished using cellular mRNA and **reverse transcriptase**. The single stranded DNA copies of mRNA are then converted into double stranded DNA by DNA polymerase, and then introduced into cloning vectors as before. Each clone of such a cDNA library is likely to contain the protein coding segments of the genome. There are distinct advantages of a cDNA library that can be exploited for certain needs. For example, a cDNA library represents the uninterrupted coding sequence of a gene, and depending upon the source of mRNA used as template for making the cDNA library, it would reflect the abundance of a particular mRNA at a particular stage of cell growth and development.

Perhaps the most critical part of DNA cloning is the identification of the correct recombinant by **gene screening** (F). The techniques most frequently used take advantage of the exquisite specificity of the base-pairing interactions between complementary molecules (DNA:RNA or RNA:RNA). If a small amount of the protein of interest is available, it can be used to determine the amino acid sequence of the first 20 or so amino acids. Using the genetic code, it is then possible to synthesize radioactively labeled DNA oligomers which can be used as nucleic acid probes (D), to search for bacterial colonies containing the gene of choice.

An important method to confirm the biological relevance of protein encoded by the cloned cDNA is to "over express" the protein product. **Expression vectors** (E) designed for this purpose have been derived from plasmid vectors (B) or viral DNA sequences which contain strong signals for the expression of cDNA. By such genetic engineering, bacteria, yeast or tissue culture cells can be induced to produce vast quantities of useful proteins such as Factor VIII, human growth hormone, interferons and viral antigens for vaccine production. It is also possible to incorporate specific **reporter genes** (H) for enzymes such as chloramphenicol acetyltransferase (CAT) or luciferase, into these expression plasmids. A reporter gene is a specific gene whose product can be assayed easily and whose levels of expression under the control of particular promoter elements is taken as an indication of the efficiency or strength of the promoter/enhancer sequences.

Items 145-148

(A)   Introns
(B)   Exons
(C)   Both
(D)   Neither

145.   Contain nucleotide sequences read by RNA polymerase II.

146.   Contain dispensable nucleotide sequences that are not spliced into mRNA.

147.   Contain nucleotide sequences that are removed from the transcript before it is converted into mRNA.

148.   Contain nucleotide sequences that are sometimes alternatively spliced into mRNA.

ANSWERS AND TUTORIAL ON ITEMS 145-148

The answers are: **145-C; 146-A; 147-A; 148-B**. Eukaryotic genes encoding proteins are transcribed by RNA polymerase II as pre-mRNA which contains intervening sequences (**introns**) that interrupt protein coding sequences (**exons**). Although the intron sequences are largely genetic "junk" and are mostly dispensable, they must still be removed precisely to provide an error free reading frame in the resulting mRNA molecule. The process of cutting out the introns and correctly splicing the exons is an essential part of creating functional mRNA. The essential steps of mRNA splicing are universal. However, different cell types contain different protein or RNA factors that can induce alternate choice of exons during the mRNA splicing process and thus confer a unique potential for functional diversity of genes.

Items 149-153

A twelve year-old female of Mediterranean background presents with severe anemia, weight loss and general fatigue. There is also a strong family history of severe anemia. Laboratory analysis shows a severe depletion of the β-globin component of hemoglobin.

149.   The most likely diagnosis for this child is

(A)   sickle cell anemia
(B)   pernicious anemia
(C)   vitamin B_{12} deficiency
(D)   thalassemia
(E)   hepatosplenomegaly

150.   The most likely cause of this disease is

(A)   single amino acid substitution in hemoglobin α-chain
(B)   single amino acid substitution in hemoglobin β-chain
(C)   defective heme biosynthesis
(D)   excessive destruction of red blood cells
(E)   decreased β-globin synthesis

151.   The mutation in this disease is located in the

(A)   exon
(B)   intron
(C)   5'-splice junction
(D)   3'-splice junction
(E)   UsnRNA

152. The primary mechanism of this mutation involves
    (A)   selection of cryptic 5'-splice site
    (B)   cryptic 3'-splice site
    (C)   blockage of mRNA transport
    (D)   selective degradation of β-globin mRNA
    (E)   translation of abnormal β-globin

153. Which technique is most appropriate for detection of the fundamental mutation in this disease?

    (A)   Western blot
    (B)   RNAse protection assay
    (C)   restriction endonuclease fragment length polymorphisms
    (D)   S1 nuclease protection assay
    (E)   nucleotide sequencing of entire exon

ANSWERS AND TUTORIAL ON ITEMS 149-153

The answers are: **149-D; 150-E; 151-C; 152-A; 153-D**. **Thalassemias**, of which there are two major subtypes, α or β, are characterized by depletion of either **α or β globins**. Two molecules each of α and β globin are required for the formation of normal functional **hemoglobin**. Decreased synthesis of either globin polypeptide chain can lead to severe and debilitating anemias.

Both α and β globin genes are transcribed as larger pre-mRNA molecules with intervening sequences (introns) IVS 1 and IVS 2 that interrupt the amino acid coding sequences. The IVSs are removed by cleavage at their 5' and 3' borders and the protein coding sequences (called exons), spliced to give rise to the contiguous protein coding sequences in the globin mRNA.

The signals crucial to the splicing of mRNAs reside directly at the splice junctions themselves. Introns always seem to begin with GU (GT in DNA) and end with AG at their 3' ends. The consensus sequence for 3' splice junctions is composed of a pyrimidine-rich region of variable length (at least ten nucleotides long), followed by a short consensus sequence ending in AG at the 3' boundary of introns. The importance of the **consensus splice junction sequences** is underscored in mutations of human globin genes that result in thalassemias.

Clinical tests show that the synthesis of globin polypeptides is defective in thalassemias. DNA sequencing has traced a number of thalassemias to base substitutions that interfere with the correct excision of the IVSs from the globin pre-mRNAs. In some types of β-thalassemia, a substitution of G to A results in the mutation of the invariant GU to AU at the 5' end of intron 1. Messenger RNA splicing is not prevented as a result of this mutation, rather, a new splice site (a cryptic splicing site which is not commonly used), becomes activated. So although globin mRNA is formed, its protein coding sequences are defective downstream from the cryptic splice site, resulting in a nonfunctional or incomplete globin polypeptide chain. Other thalassemias result from mutations that create a new 5′ or 3′ splice junction. In each case, defective mRNA is formed with out-of-frame globin coding sequences which yield non-functional polypeptides. In spite of the presence of normal α-globin chains, defective or incomplete β-globin results in non-functional hemoglobin and anemia.

The 5′ ends of the thalassemia β-globin mRNA (that arose by the cryptic splicing process), are detected by an **S1 nuclease protection assay**. In this technique, a radioactive single-stranded DNA segment from the 5′ end of the gene corresponding to the globin mRNA is annealed to the RNA isolated from the thalassemic patient. Digestion with S1 nuclease (which removes only single stranded DNA or RNA but spares RNA-DNA duplex structures), leaves only the portion of the radioactive DNA that forms a perfect hybrid with the globin mRNA. Analysis of the S1 protected radioactive DNA in denaturing sequencing gels reveals the location of the 5′ end of thalassemic mRNA.

Examine the transmission electron micrograph in **Figure 2.2** below and then choose the best response to the items below.

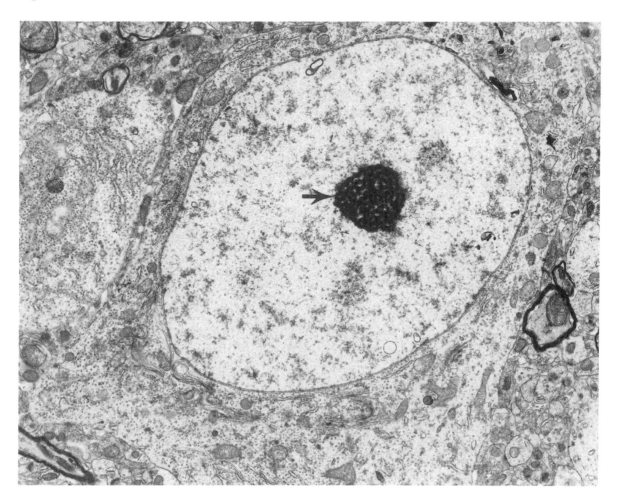

Figure 2.2

154. The structure at the pointer is a

   (A)   nucleosome
   (B)   kinetochore
   (C)   nucleolus
   (D)   nuclear envelope
   (E)   heterochromatin mass

155. The chief function of this structure is

 (A) ribosomal RNA synthesis
 (B) storage of inactive DNA
 (C) transport of mRNA out of the nucleus
 (D) boundary between nucleus and cytoplasm
 (E) attachment site for spindle microtubules

156. This structure is most obvious during which portion of the cell cycle?

 (A) interphase
 (B) mitosis
 (C) prophase
 (D) telophase
 (E) anaphase

157. This structure would be most metabolically active in which type of cell?

 (A) one actively engaged in phagocytosis
 (B) one actively engaged in steroid biosynthesis
 (C) one actively engaged in protein synthesis
 (D) one actively engaged in contraction
 (E) an erythrocyte

158. Variations in size of this organelle are primarily due to variation in the

 (A) rate of DNA synthesis
 (B) amount of granular component
 (C) amount of fibrous component
 (D) amount of histone turnover
 (E) rate of cell motility

ANSWERS AND TUTORIAL ON ITEMS 154-158

The answers are: **154-C; 155-A; 156-A; 157-C; 158-B. Figure 2.2** is a transmission electron micrograph of a neuronal cell body in the central nervous system. The arrow points to the **nucleolus**. In interphase nuclei, especially in cells actively engaged in protein synthesis, the nucleolus is often a prominent feature easily seen in the light microscope. Electron microscopy of the nucleolus reveals a central region with fibrous DNA and a granular component which consists of recently transcribed rRNA being assembled into ribosomes by combination with ribosomal proteins. The variability in the size of the nucleolus is primarily due to variations in the amount of granular component, a reflection of variations in rate of assembly of ribosomes.

The assembled ribosomes are then transported through the nuclear pores and into the cytoplasm where they either remain free in the cytoplasm for the synthesis of proteins for internal utilization (e.g., hemoglobin) or become bound to membranes to form the rough endoplasmic reticulum for the synthesis of secretory proteins (e.g., digestive enzymes).

Items 159-166

Match the nuclear protein in the list below with the most appropriate description of its function in the items below. Answers may be used once, more than once, or not at all.

(A)   Histone H2A
(B)   Histone H1
(C)   Cyclin
(D)   p34 kinase
(E)   Nucleoplasmin
(F)   Lamin
(G)   Lamin kinase
(H)   Lamin phosphatase

159.   level constant but activity fluctuates during cell cycle

160.   level fluctuates during cell cycle

161.   links nucleosomes together

162.   important constituent of nucleosome octamer

163.   regulates transport across nuclear pores

164.   This fibrillar protein binds peripheral chromatin to the inner surface of the nuclear envelope.

165.   regulates breakdown of nuclear envelope prior to metaphase

166.   regulates reformation of nuclear envelope during telophase

ANSWERS AND TUTORIAL ON ITEMS 159-166

The answers are: **159-D; 160-C; 161-B; 162-A; 163-E; 164-F; 165-G; 166-H. Histones** have an important role in organizing nuclear DNA into chromatin. At present, chromatin is thought to be arranged like beads on a string, with the beads formed by an octamer of histones [2 H2A (A), 2 H2B, 2 H3 and 2 H4] and the string being DNA. H1 (B) binds to the nucleosome and also self associates and thus functions in chromatin packing.

Cyclin (C) exists in two forms, a mitotic cyclin and a G^1 cyclin. Both are thought to play an important role in regulation of the cell cycle by causing activation of a **p34 kinase** (D). Mitotic cyclin increases rapidly prior to M and activates p34 kinase which exists at a steady state level during the cell cycle. Activation of p34 kinase in turn causes phosphorylation of specific target proteins which in turn trigger mitosis. Next, mitotic cyclin is degraded, p34 kinase activity falls and the cycle begins again. A similar regulatory cycle exists where G^1 cyclin activates p34 kinase which in turn phosphorylates target proteins. These phosphorylated target proteins then trigger DNA synthesis. Next, the G^1 cyclin is degraded.

Nucleoplasmin (E) is a pentameric complex that is involved in selective transport of regulatory proteins from the cytoplasm into the nucleus through nuclear pores. Nucleoplasmin has a short amino acid sequence on the -COOH terminal that is recognized by a receptor in the nuclear pore complex. Binding via this short sequence is a prerequisite for transport.

The inner layer of the nuclear envelope has a thin coat of 10 nm fibrils made up of **lamin** (F). There are three types of lamin (A, B, and C). They are related to intermediate filament proteins such as keratin and vimentin. Their function is probably to bind chromatin to the inner face of the nuclear envelope. **Lamin kinase** (G) phosphorylates lamins and participates in the control of the breakdown of the nuclear envelope prior to metaphase. **Lamin phosphatase** (H) dephosphorylates phosphorylated lamin. It is involved in regulation of reformation of the nuclear envelope during telophase.

Examine the electron micrograph of a freeze-fracture-etch specimen in **Figure 2.3** below and then choose the best response to the items below.

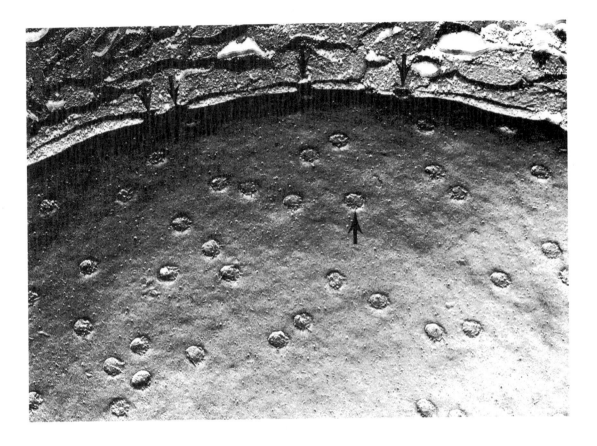

Figure 2.3

167.   The structures at the arrows are

   (A)   nuclear pore complexes
   (B)   coated pits
   (C)   gap junctions
   (D)   ribosomes on endoplasmic reticulum
   (E)   electron transport particles

168. These structures have which important function in the cell?

   (A) protein synthesis for export
   (B) protein synthesis for internal utilization
   (C) endocytosis
   (D) regulation of transit across the nuclear envelope
   (E) pinocytosis

169. The predominant protein complexes of this structure are arranged with what sort of symmetry?

   (A) cubic
   (B) pentagonal
   (C) hexagonal
   (D) octagonal
   (E) none

170. Which recently discovered protein is involved in regulation of transit through these structures?

   (A) clathrin
   (B) lamin
   (C) cyclin
   (D) nucleoplasmin
   (E) gelsolin

171. What effect would cleavage of the -COOH terminal of this protein have on receptor function?

   (A) prevents binding and transport
   (B) prevents digestion
   (C) prevents endocytosis
   (D) prevents retrograde transport
   (E) facilitates binding and transport

ANSWERS AND TUTORIAL ON ITEMS 167-171

The answers are: **167-A; 168-D; 169-D; 170-D; 171-A**. The structures shown in **Figure 2.3** are **nuclear pores**. They are large (100 nm outside diameter) fenestrations in the double unit membrane nuclear envelope. Large macromolecules move from the cytoplasm to the nucleus (regulatory proteins) and from the nucleus to the cytoplasm (mRNA and ribosomes) through these channels (10 nm inside diameter).

The nuclear pore complex consists of an octamer of subunits arranged around an aqueous channel. Ions and small molecules pass freely though these large channels. Small proteins (MW ≤ 60,000) also pass rapidly through these pores. Some larger proteins are selectively transported across the nuclear pore complex. This specific transport is mediated by a large (MW = 165,000) pentameric protein called **nucleoplasmin**. Nucleoplasmin has a short amino acid sequence on the -COOH terminal that is recognized by a receptor in the nuclear pore complex. Binding via this short sequence is a prerequisite for transport. Limited proteolysis of nucleoplasmin to remove the -COOH terminal prevents binding and therefore reduces transport of the complex. This transport phenomenon is ATP-dependent, although binding of the nucleoplasmin complex is ATP-independent. In addition to having a nucleoplasmin receptor, there are probably distinct receptor functions for recognition of other proteins and protein/RNA complexes.

Items 172-175

This set of items refers to the synthesis and processing of RNA. For each numbered item choose the **ONE** lettered answer that is most appropriate.

172. Select the cellular region most likely to be the site of synthesis of messenger RNA (mRNA).

    (A)   nuclear envelope
    (B)   nucleolus
    (C)   euchromatin
    (D)   heterochromatin
    (E)   centrosome

173. Each mRNA that is actually translated into protein by ribosomes

    (A)   is much longer than the DNA coding region because the final mRNA is spliced together from a set of RNA transcripts.
    (B)   is read by ribosomes that travel along the mRNA from the 3′ poly-A tail to the 5′ GTP leader segment.
    (C)   can only be made by ribosomes that were synthesized as part of the precursor mRNA complex in the nucleolus.
    (D)   must be bound to the large ribosomal subunit by the promoter - a portion of the mRNA that facilitates assembly of the ribosome.
    (E)   consists of a series of exons, segments of RNA that were spliced together in the nucleus as intervening regions of the RNA (introns) were eliminated subsequent to transcription in the nucleus.

174. The promoter is

   (A)   a portion of the DNA that begins approximately 100 nucleotides downstream from the 3' end of the mRNA coding region.
   (B)   a complex of two or more 5 Kd proteins that facilitate attachment of RNA polymerase to the DNA coding region.
   (C)   upstream from the 5' coding region of the DNA.
   (D)   found only on genes that code for mRNAs whose affinity for ribosomes is relatively weak.
   (E)   a portion of the gene required to promote detachment of the RNA polymerase after gene transcription is completed.

175. Which of the following is the **LEAST** accurate statement about transcription factors?

   (A)   They often form dimers that bind near the promoter region.
   (B)   Some of them contain zinc fingers, which help them interact with DNA.
   (C)   They recognize GC rich regions of the DNA.
   (D)   Some of them recognize and bind to TATA boxes.
   (E)   They are homodimers of 5 Kd proteins that activate RNA polymerase.

ANSWERS AND TUTORIAL ON ITEMS 172-175

The answers are: **172-C; 173-E; 174-C; 175-E.** The amino acid sequence of most cellular proteins is specified as a series of **codons** (triplets of nucleotides) carried by nuclear DNA, transcribed into **complementary codons** in mRNA and translated by ribosomes reading mRNAs to achieve protein synthesis. The synthesis of mRNA is carried out by **RNA polymerase II** transcribing that portion of a gene that codes the primary mRNA transcript. Most such genes are found in euchromatic regions of the nucleus. The primary gene transcript is an RNA that is much longer than the final mRNA. In addition to regions (**exons**) that actually code for the protein product of the specific protein, such primary transcripts contain interspersed segments (**introns**) that do not code or the protein; introns are removed from the primary transcript in a process that causes the exons to be joined (spliced) to one another to form the mRNA whose combined coding sequences will be translated by ribosomes into the protein specified by the gene. Although RNA transcription by RNA polymerase II begins at the start codon, binding of the polymerase to the gene actually occurs at a region of the DNA - called the **promoter** - that is many nucleotides upstream (before the 5' end) from the start codon. Binding to the promoter is essential to assure that RNA transcription occurs with high efficiency and fidelity. Proper binding at the promoter requires a number of proteins - **transcription factors** - that function to assure binding of the polymerase to the promoter. Transcription factors are a heterogeneous group of proteins that share a number of features related to their ability to bind reversibly to regulatory portions of the gene; these include the fact that they contain segments called **zinc fingers**, that they recognize and bind to **TATA**

boxes and/or GC rich regions, and that they often form dimers that bind to both strands of the DNA near the promoter region.

Items 176-179

The following set of items refers to the cytoplasmic localization of mRNA molecules.

176.   Messenger RNA (mRNA) is **LEAST** likely to be found

   (A)   in the nucleus
   (B)   traversing a nuclear pore
   (C)   associated with ribosomes of the rough endoplasmic reticulum
   (D)   in the cytoplasm but devoid of any associated proteins
   (E)   interacting with small ribosomal subunits and initiation factors

177.   The β isotype of actin is most likely to be found

   (A)   in the nucleolus
   (B)   forming the thin filaments of skeletal muscle
   (C)   at the base (proximal margin) of lamellipodia in migrating fibroblasts
   (D)   within the bilayer of the plasma membrane
   (E)   associated with desmosomal contacts between plasma cells

178.   The mRNA coding for β actin is most likely to be found

   (A)   in the nucleolus
   (B)   attached to ribosomes on the outer membrane of the rough endoplasmic reticulum
   (C)   at the base (proximal margin) of lamellipodia in migrating fibroblasts
   (D)   complexed with the ribosomes of stress fibers
   (E)   in tertiary lysosomes

179.   The β actin mRNA localization seen above (Item 178) is most likely due to

   (A)   a signal sequence at the amino terminus of the β actin polypeptide
   (B)   the TATA box upstream of the promoter region of the DNA coding for β actin
   (C)   covalent bonds formed between the mRNA and the nascent polypeptide chain
   (D)   a specific portion of the 3′ untranslated region (UTR) of the β actin mRNA
   (E)   the highly basophilic nature of the plasma membrane at certain places in cells

ANSWERS AND TUTORIAL ON ITEMS 176-179

The answers are: **176-D; 177-C; 178-C; 179-D**. Current research indicates that mRNAs are rarely, if ever, devoid of an associated coating of specific proteins related to its synthesis (in the nucleus), or its passage (through nuclear pores) to the cytoplasm, or its association with the initiation complex, or its assembly into functioning polysomes. In addition to its "known" localizations, it has recently become clear that mRNAs may be localized to specific cytoplasmic regions that correspond to the localization of the gene product for which the mRNA codes. One such case is the mRNA for β actin. This isoform is found at highest concentration in the peripheral cytoplasm of migrating fibroblasts at the areas from which lamellipodia extend. In addition to localization of the β actin molecule in these areas, it has been found that the mRNA coding for the β actin is **also** localized in such areas. A series of elegant experiments involving the construction of genetically altered genes and transcripts has shown that this localization of the specific mRNA is not dependent on translation of the mRNA but rather to the sequence of nucleotides in one portion of the untranslated 3′ tail region (3′ UTR) of the mRNA.

Items 180-182

  (A)   Monozygotic twins
  (B)   Dizygotic Twins
  (C)   Both
  (D)   Neither

180.   can be same sex or different sex

181.   are invariably genetically and phenotypically identical

182.   when of the same sex, these twins always will be indistinguishable

ANSWERS AND TUTORIAL ON ITEMS 180-182

The answers are: **180-B; 181-D; 182-D. Monozygotic twins** are derived from one zygote, i.e., they are formed from one zygote and therefore one egg and one sperm. The zygote then divides into two, assuring that the twins are usually very similar. As a result, monozygotic twins will invariably be of the same sex and share many common genetic markers.

It is possible that discordance can arise in monozygotic twins. For example, if the blood supplies of organs vary during development, they might turn out to be discordant for structures dependent upon the variant blood supplies. Somatic mutations or chromosomal rearrangements occurring after the zygote has formed might also lead to discordance. Monozygotic females might also end up genetically distinct due to preferential expression of maternal X-linked genes rather than paternal X-linked genes or as a result of postzygotic chromosomal mosaicism.

Dizygotic twins are derived from separate sperm/egg combinations and are therefore no more genetically similar than normal siblings. They can be of the same sex or different sexes. They will have many differences in their genetic markers. In rare instances, they can even be derived from different fathers.

CHAPTER III
CYTOSKELETON AND CELL MOTILITY

Items 183-186

Examine the electron micrograph of a longitudinal section of skeletal muscle in **Figure 3.1** below and then match the lettered structure on the micrograph with the most appropriate description of this structure in the items below.

Figure 3.1

183. This structure contains an enzyme whose ability to release ADP and P_i is the rate-limiting step in muscle contraction.

184. The protein actin is the principal component of this structure.

185. When Ca^{2+} binds to troponin here, it facilitates cyclic interactions between myosin and actin.

186. Glucose-6-phosphate is produced from material stored here.

Items 187-191

Choose the lettered protein component of skeletal muscle that is the most appropriate match with the numbered item below the list of answers.

(A) α-actinin
(B) Tropomyosin
(C) Troponin-T
(D) Actin
(E) Myosin
(F) Calsequestrin
(G) Troponin-C
(H) C-protein
(I) Dystrophin
(J) Troponin-I

187. This protein serves to anchor the ends of thin filaments at the Z-line.

188. This protein is a rod-like molecule, consisting of two nearly identical α-helical subunits. It binds to 7 actin monomers within a thin filament.

189. Thick filaments are held together at the M-line via this protein.

190. This protein shows extensive sequence homology to calmodulin.

191. This protein has a subunit MW = 42,000 and binding sites for ATP, myosin, tropomyosin and troponin.

Choose the best response.

192.   The coupling of nerve cell excitation to muscle fiber contraction involves all of the following events **EXCEPT**:

    (A)   release of acetylcholine from vesicles that fuse with the presynaptic membrane
    (B)   binding of acetylcholine to receptors
    (C)   hydrolysis of acetylcholine by tropomyosin
    (D)   opening of ion channels within the sarcolemma
    (E)   conduction of an action potential along t-tubule membranes

193.   When skeletal muscle is stimulated to contract, which event occurs first?

    (A)   troponin C binds Ca^{2+}
    (B)   myosin cross bridges attach to actin
    (C)   actin hydrolyses ADP
    (D)   tropomyosin dissociates from myosin
    (E) Ca^{2+} diffuses out of mitochondria

194.   When skeletal muscle relaxes following contraction, all of the following occur **EXCEPT**:

    (A)   release of myosin molecules from thick filaments
    (B)   reduction of free Ca^{2+} concentration in the sarcoplasm
    (C)   uptake of Ca^{2+} by the sarcoplasmic reticulum
    (D)   synthesis of phosphocreatine from creatine and ATP
    (E)   release of lactate into the bloodstream

ANSWERS AND TUTORIAL ON ITEMS 183-194

The answers are: **183-B; 184-A; 185-A; 186-E; 187-A; 188-B; 189-H; 190-G; 191-D; 192-C; 193-A; 194-A. Figure 3.1** shows that the contractile apparatus of **skeletal muscle** is organized in cylindrical structures called **myofibrils**. These consist of interdigitating arrays of **thick** (B) and **thin** (A) **filaments**. The principal component of thick filaments is **myosin**, a protein consisting of two heavy chains (MW ≈ 200,000) and four smaller light chains. Myosin molecules have a rod-like tail that associates with other myosin tails to form the shaft of the thick filaments. The "heads" of myosin molecules protrude from the shaft of thick filaments forming cross bridges that transiently associate with the **actin** of the thin filaments. The letters C, D and E label Z-lines, M-line, and glycogen granules, respectively.

Actin (D) has a subunit MW = 42,000. The globular (G-) actin subunits bind ATP. When G-actin polymerizes to form filamentous (F-) actin, the **bound ATP is hydrolysed to ADP**. While this hydrolysis is important in the functioning of non-muscle actins, it appears to play no role in muscle contraction. In addition to the nucleotide binding site, actin has binding sites for myosin and a number of other proteins. The myosin heads bind ATP and rapidly hydrolyse it to ADP and inorganic phosphate (P_i). The energy liberated by splitting the terminal phosphate bond in the ATP is stored in an **altered configuration of myosin** (E) and is only released when myosin heads can interact with actin in the adjacent thin filaments. When myosin heads bind actin, they release the ADP and P_i and the stored energy is used to drag thin filaments past thick filaments, thus doing work as the myofibrils shorten. The sliding movement of thick and thin filaments over one another is what is referred to as the **sliding filament model** of skeletal muscle contraction. The normal functioning of myofibrils is dependent on several proteins in addition to actin and myosin. Thin filaments are anchored to Z-discs via interaction with the protein α-**actinin** (A), while **C-protein** (H) forms cross-links between thick filaments, maintaining them in register.

In resting muscle, interaction of actin and myosin is inhibited by the proteins **tropomyosin** (B) and **troponin**. Tropomyosin is a rod-like molecule made up of two nearly identical α helical polypeptides (MW = 35,000) that wrap around each other in a coiled coil. Tropomyosin binds in the groove of the F-actin core of the thin filaments; each tropomyosin makes contact with seven actin monomers. Troponin (Tn) consists of three distinct subunits. Tn-T (C) binds the complex to tropomyosin and thus to the thin filament. Tn-I (J) interacts with actin. Tn-C (G), a Ca^{2+} binding protein with 70% sequence homology to **calmodulin**, is a Ca^{2+}-activated switch. At the low levels characteristic of resting muscle, Tn-C has no bound Ca^{2+} and the Tn complex exists in a conformation that causes tropomyosin to be positioned so that it blocks the access of myosin to the active sites on actin.

Skeletal muscle contraction is stimulated when an action potential, propagating along a nerve fiber, arrives at the **neuromuscular junction** and triggers fusion of synaptic vesicles with the presynaptic nerve cell membrane. **Acetylcholine** (ACh) released from vesicles diffuses across the synaptic cleft and binds to ACh receptors clustered in the postsynaptic sarcolemma. Binding of ACh to the sarcolemma triggers the opening of **ion channels** in the membrane and propagation of an action potential over the surface of the muscle fiber as well as inward, along extensions of the sarcolemma (muscle cell plasma membrane) called **t-tubules**. The electrical signal moving along the t-tubules causes opening of Ca^{2+} channels in adjacent patches of the sarcoplasmic reticulum (SR). The Ca^{2+} that is released is then bound to Tn-C, causing a conformational change in the Tn complex that moves tropomyosin to a position no longer blocking actin-myosin interaction. During this interaction, the ADP and P_i are released and the energy stored in the myosin is used to cause sliding of the actin filament past the myosin filament. Myosin heads repeatedly bind molecules of ATP which causes dissociation of myosin from actin, hydrolyse them to ADP and P_i and, if Ca^{2+} levels remain high enough, bind to thin filaments, again releasing ADP and P_i and causing another step of movement of thin filaments past thick filaments.

When nerve stimulation ceases, release of ACh stops, residual ACh in the neuromuscular junction cleft is destroyed by **acetylcholinesterase**, the sarcolemma repolarizes, and elements of the SR regain their relatively high impermeability to Ca^{2+} ions. These ions are pumped back into the lumen of the SR where they once again become bound to the protein **calsequestrin** (F). With cytosolic free Ca^{2+} levels low, actin-myosin interaction is again blocked by the troponin-tropomyosin complex. During brief contractile activity, ATP (broken down by actomyosin) is transiently maintained at needed levels by the action of **creatine kinase**, an enzyme which catalyzes the transfer of high energy phosphate from phosphocreatine to ADP. Contractile events lasting more than a few seconds deplete phosphocreatine and ATP levels are maintained by breakdown of glycogen stores. Glucose-6-phosphate derived from glycogen is degraded anaerobically to lactate, yielding ATP. Lactate accumulates in the sarcoplasm during prolonged muscular activity. Recovery after prolonged contraction involves resynthesis of ATP and phosphocreatine. Lactate released from the muscle is carried by the bloodstream to the liver, where it is reconverted to glucose by gluconeogenesis.

Dystrophin (I) is a protein associated with the inner leaflet of the sarcolemma. It is thought to serve an important function in reinforcing the cell membrane to strengthen it during contractions. Dystrophin structure is abnormal in Duchenne's muscular dystrophy.

Match the appropriate lettered protein with the structure or function indicated by the numbered items that follow.

(A)	Desmin
(B)	Ankyrin
(C)	Profilin
(D)	Dynein
(E)	Fragmin
(F)	Actin
(G)	α-actinin
(H)	Myosin
(I)	Keratin
(J)	Kinesin
(K)	Integrin
(L)	Tubulin
(M)	Fibronectin
(N)	Spectrin

195.   This protein binds tightly to actin monomers, reducing their ability to polymerize into microfilaments.

196.   Interaction with this integral membrane protein is one means by which components of the cytoskeleton are linked to the plasma membrane.

197.   This protein is an ATPase that helps move membrane-bound organelles from the perinuclear cytoplasm of cells such as neurons to the ends of cytoplasmic extensions.

198.   Found only in red blood cells, this protein plays a major role in determining the shape and distensibility of normal red blood cells.

199.   Hereditary spherocytosis involves the failure of this cytoplasmic protein to link spectrin to the plasma membrane.

200.   It is found as heterodimers, with a total MW $\approx$ 110,000, that bind GTP and GDP.

201.   This extracellular protein is thought to affect the cytoskeleton via binding to transmembrane proteins.

202. In nemaline myopathy, this protein is overproduced and accumulates in rod-like arrays adjacent to Z-disks.

ANSWERS AND TUTORIAL ON ITEMS 195-202

The answers are: **195-C; 196-K; 197-J; 198-N; 199-B; 200-L; 201-M; 202-G**. With the exception of **integrin** (K) and **fibronectin** (M) all the proteins in this extended matching set are cytoplasmic components that are either part of, or interact with, the cytoskeleton. **Desmin** (A) is the subunit of most intermediate filaments in muscle cells, while **keratin** (I) is the primary component of intermediate filaments in epithelial cells. **Ankyrin** (B) links the protein **spectrin** (N), found only in red blood cells, to the inner surface of the red blood cell membrane. Spectrin is thought to be important for maintaining the shape of red blood cells. Ankyrin is lacking in patients with hereditary spherocytosis. **Profilin** (C), found in many cells, can bind actin monomers, thus removing them from the polymerization-competent pool and causing microfilaments to disassemble. **Gelsolin**, another actin binding protein, reduces the tendency of microfilaments to be transiently cross-linked into gel-forming arrays. **Fragmin** (E) is one member of a family of proteins that can bind to microfilaments and cause them to be broken into shorter filaments, thus "fragmenting" them. **Actin** (F) and **myosin** (H) are found not only in muscle cells but also in most non-muscle cells, where they play roles in cell motility and in stabilizing cell shape. The protein α-**actinin** (G), involved in the attachment of actin filaments to the Z-line in skeletal muscle, and overproduced in nemaline myopathy, may also play a role in some non-muscle cells attaching microfilaments to specific patches of the plasma membrane. **Kinesins** (J) are ATPases, smaller than and distinct from myosins and **dyneins** (D), that cause movement of membrane limited vesicles along microtubules, usually away from the cytocenter and out toward the periphery of the cell. Integrins (K) are a class of integral membrane proteins that have specific domains exposed both at the cytoplasmic side of the membrane and at the extracellular face. As such, they can serve as transmembrane linkers allowing information related to extracellular conditions to affect the way cytoskeletal components are arranged. Fibronectin (M) is a large extracellular protein (which can interact with specific integrins) that provides directional clues to certain motile cells. **Tubulins** (L) are the major components of the microtubules found in cilia and flagella, as well as throughout the cytoplasm of many cell types. Tubulins are heterodimers of α and β subunits; the subunits have MW ≈ 55,000, so the tubulin dimer has a MW ≈ 110,000. Microtubules form by the polymerization of tubulin dimers. The dimers bind two molecules of guanine nucleotide, with most dimers containing one molecule of GTP and one of GDP. GTP is split to yield GDP during the polymerization process.

Items 203 and 204

203. All of the following processes directly involve the binding and/or hydrolysis of ATP **EXCEPT**

   (A)   assembly of actin monomers into microfilaments
   (B)   elongation of microtubules via addition of tubulin subunits
   (C)   movement of cilia
   (D)   transport of vesicles along axonal microtubules
   (E)   dissociation of myosin cross bridges from thin filaments in cardiac muscle

204. Which of the following correctly states a difference between the way in which smooth and cardiac muscle cells contract?

   (A)   Smooth muscle cells are innervated, whereas cardiac muscles are not.
   (B)   The myosins found in cardiac muscles do not form thick filaments as readily as do those in smooth muscle.
   (C)   Cardiac muscle contains actin, but smooth muscle does not.
   (D)   Smooth muscle has a much less extensive sarcoplasmic reticulum than does cardiac muscle.
   (E)   While smooth muscle bundles contain connective tissue fibers, cardiac muscles do not.

ANSWERS AND TUTORIAL ON ITEMS 203 AND 204

The answers are: **203-B; 204-D**. ATP is the immediate energy source for the **contraction** of skeletal and cardiac muscle, as well as in smooth muscle. The myosins of smooth muscle cells form thick filaments, but much less readily than in skeletal and cardiac muscle. In each case, the binding of ATP to myosin causes dissociation of myosin cross bridges from adjacent thin filaments and is followed by the rapid hydrolysis of ATP yielding ADP and P_i, both of which remain bound to the myosin head until the head can reassociate with actin. G-actin binds ATP and this is hydrolysed to ADP and P_i when the actin monomer polymerizes to form F-actin. Neither cardiac nor smooth muscles have the extensive innervation characteristic of skeletal muscle. While contraction of all muscles is regulated by calcium levels, smooth muscle does not have the extensive sarcoplasmic reticulum found in the striated muscles. All three types of muscle contain thin filaments made of actin and, in all three, muscle cells (fibers) are held together by extracellular connective tissue fibers, mostly collagen. Enzymes, such as **dynein** and **kinesin**, which cause vesicles to move along microtubules, utilize ATP as the energy source for such movements. **Tubulins** all bind GTP (not ATP), which is hydrolysed to GDP when microtubules are formed.

74

Items 205-210

A pregnant 39 year-old female has amniocentesis carried out because of concern about possible birth defects associated with pregnancy in older women. Strong reactivity with an antibody to glial fibrillary acidic protein (GFAP) is detected.

205.　GFAP is

　　　(A)　found in neurons of the central nervous system
　　　(B)　a glycoprotein containing numerous sulfated side chains
　　　(C)　a constituent of the nuclear envelope
　　　(D)　a GTPase secreted by glial cells and active at low pH
　　　(E)　one of the proteins that form intermediate filaments

206.　The woman's obstetrician informs her that

　　　(A)　her child is likely to be born three months premature
　　　(B)　the test results suggest that she has AIDS
　　　(C)　the fetus has spina bifida
　　　(D)　a cesarean delivery will be required
　　　(E)　she will probably give birth to triplets

207.　A finding of positive immunoreactivity for which protein in a tumor mass is indicative that the tumor is of epithelial origin?

　　　(A)　keratin
　　　(B)　desmin
　　　(C)　vimentin
　　　(D)　collagen
　　　(E)　melanin

208.　Dense tangles of neurofilaments are most likely to be found in

　　　(A)　autopsy material from the brain of a victim of Alzheimer's disease
　　　(B)　benign tumors of the esophagus
　　　(C)　the lungs of patients recovering from carbon monoxide poisoning
　　　(D)　the liver of children suffering from acute lead poisoning
　　　(E)　spinal cord tissue of a woman paralyzed due to a diving accident

209. Which of the following factors is most likely to be physiologically relevant to the disassembly of intermediate filaments?

   (A)   conjugation of the filaments with glycogen
   (B)   phosphorylation of intermediate filament protein subunits
   (C)   covalent coupling to gamma tubulin sidechains.
   (D)   ATP concentrations lower than 1 mM
   (E)   stimulation of protein kinases by free calcium levels greater than 100 micromolar

210. Which of the following is LEAST likely to be a structure or function with which intermediate filaments are associated?

   (A)   attachment to desmosomes
   (B)   binding to integrins
   (C)   linkage of myosin and actin filaments.
   (D)   anchorage onto hemidesmosomes
   (E)   connection of the cytoskeleton to the plasma membrane.

ANSWERS AND TUTORIAL ON ITEMS 205-210

The answers are: **205-E; 206-C; 207-A; 208-A; 209-B; 210-C. Intermediate filaments** are components of the **cytoskeleton**. Intermediate filaments are typically 10 nm in diameter and are formed by the polymerization of one or more members of a family of polypeptides. One of the main functions of intermediate filaments appears to be the "integration" of forces - applied to the plasma membrane - throughout the entire cytoskeleton. Intermediate filaments bind tightly and specifically to desmosomes and hemidesmosomes, apparently via interactions with one or more subtypes of the transmembrane proteins called integrins. Although the disassembly of intermediate filaments at specific points in the cell cycle is not yet well understood, it is now known that phosphorylation of intermediate filament proteins favors disassembly of the filaments. The proteins that form intermediate filaments differ from tissue to tissue and these differences are used in a number of clinical tests. Epithelial tissues have intermediate filaments made from a variety of **keratins**; the finding of keratin immunoreactivity in a tumor mass implies that the tumor is derived from epithelial tissue. Similarly, **desmin, vimentin, GFAP,** and **neurofilament protein** are diagnostic of muscle cells, connective tissues, glial cells in the CNS (including astrocytes), and neuronal cells, respectively.

   Spina bifida is a developmental defect involving failure of the posterior end of the neural tube to close. In such situations, the amniotic fluid is found to contain GFAP. Among the many defects found in patients with **Alzheimer's disease** is the presence of numerous large aggregates ("tangles") of neurofilaments.

Items 211-214

A 29 year-old Caucasian male requests treatment for infertility. Routine patient history reveals frequent colds and sinus infections. Examination of a semen sample shows large numbers of non-motile spermatozoa. Chest X-rays reveal situs inversus, including right-sided location of the heart.

211.   The most likely diagnosis is

    (A)   Down syndrome
    (B)   cryptorchism
    (C)   Kartagener's syndrome
    (D)   bronchial pneumonia
    (E)   Klinefelter's syndrome

212.   If electron microscopic studies of the patient's spermatozoa were carried out you would be most likely to find

    (A)   absence of basal bodies
    (B)   lack of mitochondria in the midpiece
    (C)   dynein arms missing from the outer doublets
    (D)   radial spokes are incorrectly coiled
    (E)   cross-links between axoneme and plasma membrane are shortened

213.   Bronchial clearance is most likely reduced because

    (A)   bronchial lumina are plugged with macrophages
    (B)   secretion by goblet cells is reduced
    (C)   cells of the lining epithelia lack all ATPase activity
    (D)   cilia on lining epithelial cells are non-motile
    (E)   the mucus produced by the lining epithelial cells is not soluble at physiological ionic strength

214.   Which of the following conditions would the patient be **LEAST** likely to have?

    (A)   a poor sense of smell
    (B)   problems with balance
    (C)   frequent need to expectorate
    (D)   heightened sensitivity to painful stimuli applied to the skin
    (E)   early hearing loss

ANSWERS AND TUTORIAL ON ITEMS 211-214

The answers are: **211-C; 212-C; 213-D; 214-D**. The patient suffers from **Kartagener's syndrome** (or triad). This is an inherited defect (autosomal) related to the absence of one or both **dynein arms** from the outer doublet microtubules of the axonemes of cilia and flagella. Dyneins are microtubule associated ATPases found attached to the outer doublets of cilia and flagella. They provide the energy (via hydrolysis of ATP) to drive the bending activities of the flagella which move spermatozoa and the cilia which cause fluid to move over certain epithelial surfaces. In individuals who lack dyneins, cilia and spermatozoa are inactive. This results in infertility in males and reduced fertility in females. It also leads to respiratory infections because of inadequate clearance of mucous secretions that trap pathogens. The combination of reduced fertility and reduced clearance of mucous secretions constitutes the immotile cilia syndrome. When situs inversus occurs (due to failed rotation of developing organs) the complete Kartagener's syndrome results. Patients with this condition often show functional defects in other organs whose activities involve cilia or modified cilia; these include olfactory epithelia, the organ of Corti, the vestibular apparatus and cells lining the ventricles of the brain. Perception of painful stimuli to the skin involves free nerve endings, structures that do not contain cilia.

Items 215-217

For each item, choose the most appropriate answer.

215.   The axoneme of a cilium

(A)   has 9 single microtubules arranged around a central pair of doublet microtubules
(B)   contains several ATPases, including myosin
(C)   is very similar to the axoneme of a sperm flagellum
(D)   shortens by about 25% when stimulated by Ca^{2+} ions
(E)   normally is activated by binding of acetylcholine at the tip of the cilium

216.   The peripheral microtubules in a cilium are

(A)   about 10 nm in diameter
(B)   constructed primarily from gamma-tubulin
(C)   stably connected to one another by dynein cross-links
(D)   disassembled when Ca^{2+} levels are sufficiently elevated to bind to calmodulin
(E)   linked to the central pair of microtubules by structures called radial spokes

217.	Dyneins

	(A)	are ATPases with extensive sequence homology to myosins
	(B)	are GTPases that bind transiently to the central pair of axonemal microtubules
	(C)	require association with actin filaments in order to effect undulatory motion of cilia
	(D)	bind to the plasma membrane surrounding the axoneme
	(E)	go through cycles of attachment to and detachment from microtubules

ANSWERS AND TUTORIAL ON ITEMS 215-217

The answers are: **215-C; 216-E; 217-E**. The movements of eucaryotic cilia and flagella are brought about by a cytoskeletal complex called the **axoneme**. The axonemes of cilia and flagella are very similar. They are cylindrical structures with a central pair of single microtubules surrounded by nine **doublet microtubules**. This organization of the axonemal complex is referred to as the "9+2 arrangement". Both the central singlets and the peripheral doublets are assembled from α-β tubulin dimers similar to those that form cytoplasmic microtubules; however, in contrast to cytoplasmic microtubules (which undergo various cycles of assembly/disassembly), the microtubules of the axoneme are stably assembled for the "life" of the axoneme. As with other microtubules, those of the axoneme are approximately 24 nm in diameter. The outer doublets are joined tightly together by links made of a protein called **nexin** and are firmly connected to a proteinaceous sheath surrounding the central pair by protein structures called **radial spokes** (or radial links). The outer doublets have associated with them multiple copies of a high molecular weight protein complex called **dynein**. Dyneins are much larger than, and distinct from the muscle enzyme myosin (cilia and flagella contain neither myosin nor actin), but - like myosins - they hydrolyze ATP and the energy liberated by that hydrolysis is used to power the bending motions of cilia and flagella. It appears that activation of cilia and flagella involves some sort of signal transmission, perhaps via the basal body, to the base of the axoneme. This activates the dynein side arms which hydrolyze ATP and transiently bind to the doublet microtubules adjacent to the doublet to which they are permanently attached. This transient interaction between two adjacent doublets causes a sliding of one doublet relative to the other; the dynein-driven sliding motions cause the axoneme to bend. It is the bending movements of the axoneme that cause fluid to move past cilia and cause flagella to propel cells through fluid. Neither cilia nor flagella shorten to any significant extent during the cyclic association and dissociation of dynein bridges with adjacent microtubule doublets.

Examine the high power transmission electron micrograph in **Figure 3.2** below. It is an array of long hollow structures cut perpendicular to their long axis. The diameter of each hollow structure is 25 nm.

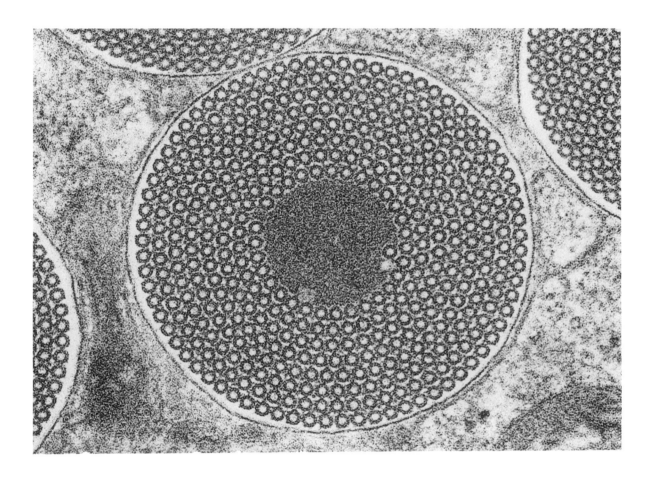

Figure 3.2

218. Each individual hollow structure is a

    (A)   microfilament
    (B)   ciliary axoneme
    (C)   microtubule
    (D)   thick filament of muscle
    (E)   intermediate filament

219. The most abundant protein found in these structures is

    (A)   tubulin
    (B)   actin
    (C)   keratin
    (D)   desmin
    (E)   myosin

220. Which of the following proteins is also likely to be present in abundance?

    (A)   actin
    (B)   vimentin
    (C)   dynein
    (D) α-actinin
    (E)   tropomyosin

221. Which description is most appropriate for these structures?

    (A)   The thick filaments are composed of aggregates of myosin molecules.
    (B)   Myosin side-arms are involved in their movement.
    (C)   They consist of globular actin molecules arranged in two intertwined helices with troponin and tropomyosin molecules arrayed along the helix.
    (D)   They consist of 13 protofilaments composed of alternating α- and β-tubulin subunits arranged like a string of beads.
    (E)   They are abundant in the cores of microvilli.

ANSWERS AND TUTORIAL ON ITEMS 218-221

The answers are: **218-C; 219-A; 220-C; 221-D**. **Figure 3.2** is a high power transmission electron micrograph of an array of **cytoplasmic microtubules** cut perpendicular to their long axis. Microtubules are long hollow structures with an outside diameter of 25 nm. They consist of 13 **protofilaments**. Each protofilament consists of alternating subunits of α- and β-**tubulin**. The tubulins have similar molecular weights (55 kD) but have different amino acid sequences and subtle functional differences. In addition to tubulin, microtubules have a variety of **microtubule-associated proteins** (MAPs) that are involved in the interaction of microtubules with other cytoskeletal elements; one of the most important MAPs is **dynein**, the ATPase responsible for the sliding of axonemal outer doublets that result in movements of cilia and flagella. Microtubules are important cytoskeletal structures that function in maintenance of cell morphology. They are also major constituents of cilia, flagella, centrioles and the mitotic spindle. Microtubules play a central role in the movements of entire cells as well as in such intracellular movements as the redistribution of chromosomes during mitosis.

Examine the high power transmission electron micrograph of cell surface projections in **Figure 3.3** below. These projections are cut perpendicular to their long axis. For each numbered item of a functional role choose the most appropriate component labelled with a letter in the electron micrograph.

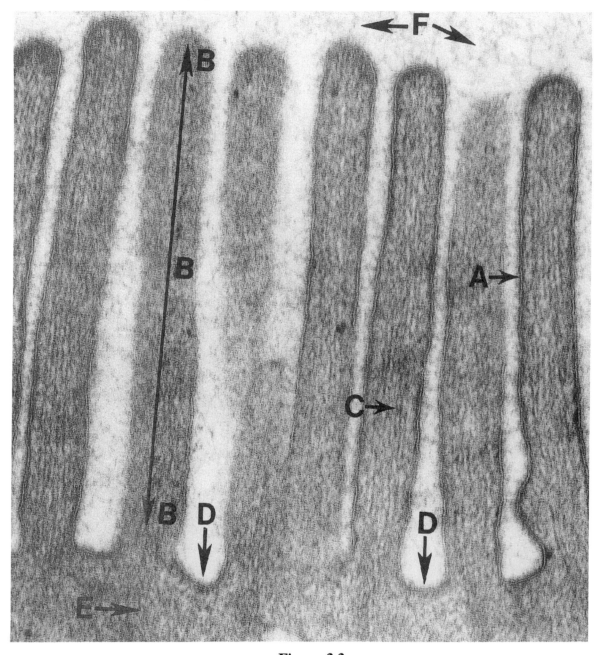

Figure 3.3

222.   These structures are actin-rich microfilaments. They serve to stiffen the surface
        projections.

223.   This structure is a phospholipoprotein bilayer serving as a selective permeability
        barrier surrounding the projections as well as the entire cell.

224.   This is a glycoprotein-rich layer containing enzymes that hydrolyse disaccharides.

225.   The projections shown here are called

        (A)   cilia
        (B)   microvilli
        (C)   dendrites
        (D)   spindle fibers
        (E)   flagella

226.   These surface projections have which primary physiological function?

        (A)   movement of luminal contents
        (B)   transduction of mechanical into electrical energy
        (C)   transduction of chemical into electrical energy
        (D)   absorption of luminal contents
        (E)   gas exchange

ANSWERS AND TUTORIAL ON ITEMS 222-226

The answers are: **222-C; 223-A; 224-F; 225-B; 226-D. Figure 3.3** is a high power
transmission electron micrograph of sections cut perpendicular to the long axis of the
microvilli (B) in the jejunum of the small intestine. Microvilli are apical surface projections
designed to **increase the cell surface area**. They are particularly well developed in areas
where luminal contents are being absorbed, e.g., in the small and large intestine and proximal
and distal convoluted tubules in the kidneys. Like all other cell surface projections, microvilli
are covered by a phospholipoprotein bilayer of plasma membrane (A). The cores of microvilli
are filled with many **microfilaments** (C) with a diameter of 6 nm. These microfilaments are
rich in **actin** and are involved in stabilizing the microvilli. The plasma membrane of intestinal
microvilli has a thick **glycocalyx** (F) which contains the glycoprotein rich extracellular
domains of integral membrane proteins. This fuzzy coat on the outer leaflet of the plasma
membrane consists of minute filaments, 2.5-5 nm in diameter. They often project 0.1 to 0.5
μm beyond the apical tips of microvilli. The glycocalyx prevents digestive enzymes in the
lumen of the small intestine from gaining access to epithelial cells. In addition, the glycocalyx
contains digestive enzymes, such as **disaccharidases,** that complete the final steps in nutrient
digestion.

Examine the transmission electron micrograph of skeletal muscle in **Figure 3.4** below. Match the labeled structure in the micrograph with the most appropriate description of its microscopic anatomy or physiological role.

Figure 3.4

227. These structures are the ends of individual sarcomeres. They are regions rich in α-actinin.

228. These structures are regions of overlap between actin-rich thin filaments and myosin-rich thick filaments.

229. These structures produce ATP, which is hydrolyzed to provide the energy for muscle contraction.

230. These structures are regions rich in actin-containing thin filaments where there is no overlap between thin and thick filaments.

231. These structures are regions rich in myosin-containing thick filaments where there is no overlap between thick and thin filaments.

ANSWERS AND TUTORIAL ON ITEMS 227-231

The answers are: **227-B; 228-C; 229-A; 230-D; 231-E. Figure 3.4** is a high magnification transmission electron micrograph of skeletal muscle cut in longitudinal section. Skeletal muscle fibers contain many **myofibrils**, each consisting of many **sarcomeres**. Each sarcomere is formed by a regular array of thick (myosin-rich) and thin (actin-rich) filaments. Myofibrils are composed of many sarcomeres. Individual sarcomeres extend from **Z line** (B) to Z line. Sarcomeres have central **A bands** (C) where there is extensive overlap between **thin, actin-rich filaments** and **thick, myosin-rich filaments**. When a muscle fiber (cell) contracts, the width of the **I bands** (D) and **H bands** (E) decrease because the thin and thick filaments increase their overlapping due to sliding of thin filaments past thick filaments. The H bands are regions of no overlap between thin and thick filaments. H bands contain no thin filaments. The length of the A bands remains constant during muscle contraction; however, the I bands and H bands decrease in length and the Z lines move closer together, leading to a shortening of the sarcomere. When many sarcomeres shorten, the entire muscle cell shortens. Muscle contraction is driven by the energy rich compound ATP which is produced in **mitochondria** (A).

Contractile force in muscle is generated by a change in the position of **actin** and **myosin**, which is regulated by intracellular calcium concentration. Energy for muscle contraction is derived from the hydrolysis of ATP. Under appropriate conditions, the actin-myosin complex has ATPase activity. Release of energy from ATP hydrolysis causes conformational changes in the muscle proteins resulting in useful movement. A sarcomere has a variable total length depending on the contractile status of the cell. When a muscle contracts, thick and thin filaments slide past one another. During a contraction and relaxation cycle, calcium concentration around the myofibrils increases suddenly. This causes a conformational change in the **troponin** molecule, which exposes the S-1 (cross-bridge) binding site of actin, and a myosin-actin complex forms. Another conformational change occurs, and the S-1 fragment, still in association with the actin-containing thin filament, swings like an oar in an oarlock and causes the thin filament to slide relative to the thick filament. When this occurs at millions of cross-bridges, the entire sarcomere is shortened. The calcium concentration falls rapidly, following hydrolysis of ATP and the swinging of cross-bridges. The drop in calcium severs the association between actin and myosin and the contraction stops. ATP is hydrolysed to adenosine diphosphate (ADP), which subsequently is phosphorylated to form ATP. Skeletal and cardiac muscle cells have large numbers of mitochondria, which synthesize the large amounts of ATP required for the work done by muscle cells contracting against and external load.

Figure 3.5 is a freeze-fracture electron micrograph of the apical surfaces of epithelial cells lining the lumen of the ileum. For each numbered item naming, or stating the function of a specific structure, choose the **ONE** best answer.

Figure 3.5

232. The surface projections indicated by the arrows are

    (A)   microvilli
    (B)   cilia
    (C)   basal bodies
    (D)   desmosomes
    (E)   flagella

233. If these structures were isolated by cell fractionation and examined by gel electrophoresis, which of the following proteins would most likely be found in large amounts?

    (A)   dynein
    (B)   hemoglobin
    (C)   tubulin
    (D)   actin
    (E)   spectrin

234. The membranes covering these surface projections

    (A)   are held together by numerous desmosomes
    (B)   do not have a phospholipid bilayer
    (C)   are continuous with the plasma membranes of the epithelial cells
    (D)   synthesize the glycoproteins which cover the projections
    (E)   lack intramembrane particles

235. A transmission electron micrograph of the same projections would be most likely to reveal

    (A)   numerous microtubule protofilaments
    (B)   bundles of microfilaments
    (C)   thin filaments containing actin interdigitating with numerous thick filaments made of myosin
    (D)   many secretory vesicles attached to bundles of intermediate filaments of the keratin superfamily
    (E)   dense patches on the interior (cytoplasmic) surfaces of the membranes, to which small bundles of intermediate filaments attach

236. If the filaments that fill the cores of the projections were traced "downward" into the apical cytoplasm of the epithelial cells it would be seen that

    (A)   they insert into a terminal web containing actin microfilaments and a special myosin isotype
    (B)   all the filaments terminate on dense bodies comprised of the protein α-actinin
    (C)   the filaments insert into microtubule organizing centers that surround the basal bodies from which the projections radiate
    (D)   they become continuous with the intermediate filaments that anchor the projections to the Golgi apparatus
    (E)   the filaments forming the cores of the projections are oriented so that adjacent filaments have opposite polarity from one another, thus guaranteeing that filament ends at the tips of the projections are not all of the same polarity

ANSWERS AND TUTORIAL ON ITEMS 232-236

The answers are: **232-A; 233-D; 234-C; 235-B; 236-A. Figure 3.5** shows that the apical surfaces of many absorptive epithelial cells contain projections called **microvilli;** when such microvilli are very numerous - as is the case here - they are referred to as a brush border. Microvilli are surface projections whose principal function is to provide an expanded membrane surface into which transport proteins can be inserted; the membrane covering them is a typical phospholipid bilayer (with assorted integral and peripheral membrane proteins) and is continuous with the plasma membrane of the cell proper. Each microvillus is supported in its extended form by a core bundle of **actin microfilaments**. All the microfilaments in microvilli are oriented with the same polarity; their "minus" or "barbed" ends are inserted into some dense material at the tip of the projection, thus they are all oriented in the "same" polarity as is seen with the insertion of the thin filaments of skeletal muscle cells into the Z-lines of sarcomeres. Microvilli do not appear to function as motile organelles; they do not contain significant amounts of **myosin,** nor do they have other cytoskeletal proteins such as tubulins, dyneins, or intermediate filament proteins. The microfilaments of the microvillar cores extend down into the cortical (or terminal) web that fills the apical cytoplasm of the epithelial cells. The terminal web contains other actin-microfilaments, that insert into the tight junctions linking the epithelial cells together. The **cortical web** also contains a special myosin that forms small myosin filaments, whose function is still not well understood. Immediately below the cortical web, the cytoplasm contains large numbers of **intermediate filaments**; these insert into the **desmosomes** that also help hold adjacent epithelial cells together.

Figure 3.6 is a scanning electron micrograph of the apical surfaces on the epithelial cells lining the trachea. For each numbered item of identity or function, choose the **ONE** best answer.

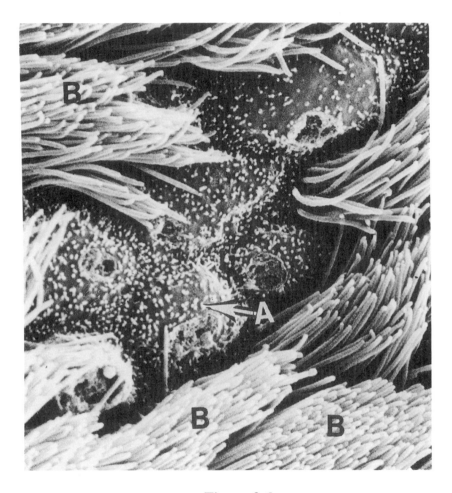

Figure 3.6

237.   The structures labelled A are

    (A)   cilia
    (B)   flagella
    (C)   microvilli
    (D)   microtubules
    (E)   collagen fibers

238.   The structures labelled B are

   (A)   cilia
   (B)   flagella
   (C)   microvilli
   (D)   microfilaments
   (E)   collagen fibers

For Items 239 through 242 use the following answers:

      (A)   Structures labelled A only
      (B)   Structures labelled B only
      (C)   Both structures A and B
      (D)   Neither structure A nor B

239.   Formed by coordinated polymerization of keratins and vimentins.

240.   Elongate structures whose core contains an elaborate array of microtubules.

241.   These structures would be expected to "collapse" upon treatment with cytochalasin B.

242.   These structures would be expected to bind rhodamine-labelled phalloidin very strongly.

ANSWERS AND TUTORIAL ON ITEMS 237-242

The answers are: **237-C; 238-A; 239-D; 240-B; 241-A; 242-A**. The scanning electron micrograph in **Figure 3.6** shows that the apical surfaces of the epithelial cells lining the trachea may have two distinct types of apical projections. The shorter ones (A) are **microvilli**; although not as numerous as the microvilli in the brush border of intestinal epithelial cells, these microvilli have a similar structure and function. They are maintained as elongate extensions of the plasma membrane by a core of actin microfilaments and serve to provide extra membrane surface to house a variety of proteins that play a role in transfer of substances across the apical surface of the epithelium. The longer projections (B) are **cilia**; they contain a core complex of microtubules called the axoneme and serve, by their coordinated "beating" to move fluids over the surfaces of the epithelium, in this case in a direction towards the lumen of the oral cavity. Because they contain a bundle of microfilaments, microvilli can be probed by agents specific for actin. They will bind the fungal toxin **phalloidin** and, if a fluorescent derivative (rhodamine-phalloidin) of that toxin is applied, the microfilaments (and thus the microvilli) will be strongly and selectively labelled. Another probe that is nearly, but not absolutely, specific for actin is the family of drugs called **cytochalasins**. Cytochalasins (the most specific of which is cytochalasin D) bind to actins in such a way as to bring about their breakdown, hence causing large scale shortening (collapse) of the microvilli.

Items 243-246

243. Microfilaments

    (A)   are 4-7 nm in diameter
    (B)   contain the proteins actin and keratin
    (C)   cause the anterograde movement of mitochondria in axons
    (D)   form the dense mass of filaments attaching to desmosomes
    (E)   bind chromosomes to the mitotic spindle

244. Actin

    (A)   is a globular protein that can assemble into filaments
    (B)   makes up most of the M-line in skeletal muscle
    (C)   causes the movements of vesicles along the microtubules of axons
    (D)   contains specific binding sites for the enzyme dynein
    (E)   can be degraded, by certain proteases, into heavy and light meromyosins

245. The actin in non-muscle cells

    (A)   is usually associated with dynein
    (B)   forms filaments that are frequently assembled and disassembled
    (C)   is usually associated with non-muscle tropomyosin
    (D)   rarely makes contact with the plasma membrane
    (E)   is usually found in sarcomeres

246. If an antibody that is specific for myosin is microinjected into a cell that has just completed DNA synthesis, which of the following is the **most** likely consequence?

(A) Cell division will be completed normally, but the daughter cells will be unable to move during the following interphase.

(B) Cytokinesis will proceed normally, but chromosome movement at the ensuing anaphase will be prevented.

(C) Injected cells will enter prophase but not proceed to metaphase because no mitotic spindle will assemble.

(D) The cell will fail to divide, but will subsequently be found to have two nuclei.

(E) There will be no discernible effect because non-muscle cells contain actin but not myosin.

ANSWERS AND TUTORIAL ON ITEMS 243-246

The answers are: **243-A; 244-A; 245- B; 246-D**. The protein **actin** is found in essentially all eucaryotic cells. In non-muscle cells actin forms filaments with a diameter of 4-7 nm, called microfilaments. In contrast to the thin filaments in muscle cells, which are fairly stable assemblies of actin, the microfilaments of non-muscle cells are dynamic structures that are frequently assembled and disassembled in response to the multiple stimuli to which non-muscle cells must respond. Most non-muscle cells also contain one or more types of **myosin**, with similarities to and differences from the myosins found in muscle cells. Actin microfilaments participate in a variety of cellular activities, many of which require the participation of non-muscle myosins. One such activity is cytokinesis, the process by which a contractile ring - containing myosin and actin microfilaments - cause a constriction at the midbody of the mitotic spindle. This constriction causes the daughter nuclei to be separated from one another into distinct cytoplasmic domains - the daughter cells. If myosin activity is disrupted by microinjection of specific antibodies the various stages of mitosis can proceed "normally" but the contractile ring fails to form and/or function properly; the result is the formation of cells with the daughter nuclei contained in a single common cytoplasm.

247. Which statement is **LEAST ACCURATE**?

   (A) The polarity of microtubules is related to the fact that the tubulin subunit is a heterodimer of α and β subunits.
   (B) Actin filaments often associate with special patches of the plasma membrane.
   (C) Kinesin is an ATPase that may interact with microtubules.
   (D) Intermediate filaments are assembled in the intermediate zone of the mitotic spindle.
   (E) While some microtubules may be rapidly assembled or disassembled, those of cilia and flagella are quite stable once assembled to form the axoneme.

248. Which of the following is most directly involved in causing axoplasmic transport?

   (A) myelin sheath
   (B) neurofilaments
   (C) synapse
   (D) neurotubules
   (E) myosin

249. Dynein and kinesin

   (A) are both active in the axoneme of cilia
   (B) hydrolyze ATP to power movement of membrane bound organelles in axons
   (C) drive the transfer of actin from areas rich in microtubules to sites where intermediate filaments form
   (D) both require troponin for activation
   (E) serve to phosphorylate microfilaments and microtubules

250. Tubulozole is a recently discovered drug that causes the depolymerization of microtubules. Which of the following is **LEAST** likely to be a **direct** result of exposure to tubulozole?

   (A) Reduction in the rate of mitosis needed to replace damaged epithelial cells in the intestinal mucosa.
   (B) Diminished rate of elongation of cilia forming on the apices of regenerating respiratory epithelium.
   (C) Functioning of the contractile ring during cell division in early embryogenesis.
   (D) Movement of secretory vesicles along axons.
   (E) Blockage of spermatogenesis at metaphase of meiosis I.

251. Of the following, the organelle **MOST DIRECTLY RESPONSIBLE** for organizing the microtubules in cells is the

(A) plasma membrane
(B) centrosome
(C) chromatin
(D) nucleus
(E) terminal web

252. The protein **MOST LIKELY** to be found in the centriole is

(A) tubulin
(B) dynein
(C) alpha actinin
(D) kinesin
(E) actin

253. Which organelle is **MOST SIMILAR** in structure and function to centrioles?

(A) microfilament
(B) desmosome
(C) axoneme
(D) basal body
(E) nuclear pore

ANSWERS AND TUTORIAL ON ITEMS 247-253

The answers are: **247-D; 248-D; 249-B; 250-C; 251-B; 252-A; 253-D. Microtubules**, made of the protein **tubulin**, are cytoskeletal components important in a large number of cellular functions. All known microtubules are formed of tubulin dimers, each made of one α- and one β-tubulin molecule. Because the dimers assemble head-to-tail (i.e., α-β-α-β), each microtubule is a polarized structure and this polarity is important in the many microtubule-based events involving directed movement. Microtubules are assembled (from the tubulin dimers) at or near structures called **microtubule organizing centers** (MTOCs). The most common MTOCs are parts of cell regions called **centrosomes**. Centrosomes are areas, usually near the nucleus (thus often called cytocenters), that contain one or two centrioles and a series of ill-defined dense areas or **satellite bodies**. It is the material of the satellite bodies that serve as MTOCs and from which microtubules elongate, usually outward toward the cell periphery. All microtubules assemble such that their "minus" end is at the MTOC and the "plus" end is near the cell periphery. This assembly polarity is important because the enzymes that move organelles along microtubule tracks are of two distinct classes.

Microtubule-based motors that move materials from the minus (cytocenter) ends to the plus (cell peripheries) ends of microtubules are **kinesins**. The transfer of materials along axons toward the axon terminus (anterograde axoplasmic transport) is powered by kinesins (ATPases) moving along neurotubules (nerve cell microtubules). Conversely, retrograde axoplasmic transport along neurotubules is powered by **dyneins**, a separate and distinct set of microtubule motors. The importance of microtubule-based cellular functions is revealed by the wide range of processes that can be inhibited by agents - such as the drug **tubulozole** - that disrupt microtubules. Of the processes listed as possible "answers" in Item 250, only the contractile ring required for cytokinesis (and made up of actin microfilaments) would **not** be directly affected by tubulozole. Although not involved in cytokinesis, microtubules **are** crucially involved in karyokinesis (the part of mitosis in which chromosomes are separated to the forming daughter cells) because they are the principal cytoskeletal component of the mitotic spindle. The spindle poles are composed of centrioles and surrounding MTOCs. Although the microtubules of the centrioles are not essential for assembly of mitotic spindle microtubules, the microtubular nature of the centrioles is presumably important to the functioning of the cellular organelles that are most closely related to centrioles - the basal bodies. Assembly of the axonemal complex of cilia and flagella requires the presence of basal bodies, organelles that appear to be identical to centrioles, but moved to the cell periphery. It is not clear what changes are effected when centrioles are repositioned to function as basal bodies, but the microtubules of ciliary and flagellar axonemes **do** assemble on the microtubules that form the wall of the basal bodies.

Refer to the transmission electron micrograph below (**Figure 3.7**).

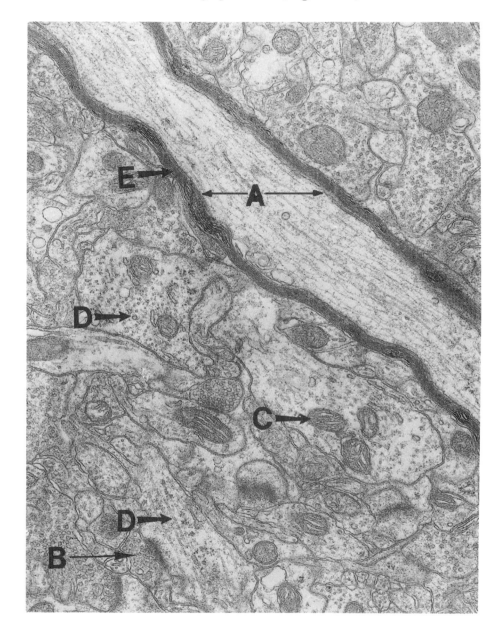

Figure 3.7

254. The elongate cellular projection labelled A is a

   (A) flagellum
   (B) nerve axon
   (C) pseudopodium
   (D) cilium
   (E) microvillus

255. Structure labelled C is

   (A) Golgi apparatus
   (B) peroxisomes
   (C) microtubules
   (D) mitochondria
   (E) secondary lysosomes

256. Structures labelled D are

   (A) Golgi apparatus
   (B) peroxisomes
   (C) microtubules
   (D) mitochondria
   (E) secondary lysosomes

257. Organelles **LEAST** likely to be found in projection A are

   (A) mitochondria
   (B) intermediate filaments
   (C) ribosomes
   (D) microtubules
   (E) membrane vesicles

258. Which of the following enzymes is **LEAST** likely to be present in projection A?

   (A) protein kinase
   (B) DNAase
   (C) kinesin
   (D) dynein
   (E) protein phosphatase

ANSWERS AND TUTORIAL ON ITEMS 254-258

The answers are: **254-B; 255-D; 256-C; 257-C; 258-B**. **Figure 3.7** is a transmission electron micrograph of neural tissue, showing a large **axon** (A) with its **myelin sheath** (B). The surrounding structures are axons and dendrites of other neurons, as well as portions of supportive - or glial - cells. One of the most prominent membranous organelles are the **mitochondria** labelled (C) which produce the ATP required to power the various synthetic and motile events occurring in these cells. Nerve axons are long slender extensions reaching from the neuronal cell bodies (not shown) to the receptor cell (also not shown) receiving a signal from the nerve cell. The activation of a neuron cell body leads to the passage of a membrane depolarization that propagates along the nerve cell plasma membrane, as an **action potential,** to the axon terminus. Axons contain a variety of organelles, but - having no ribosomes (either free or bound to rough ER) - are unable to synthesize proteins. All proteins required for the establishment and maintenance of the axon are synthesized in the nerve cell body and carried to the nerve terminus by **axoplasmic transport**.

Examine the transmission electron micrograph below (**Figure 3.8**). It shows the apical aspect of an epithelial cell with certain structures labelled. For each numbered item, select the lettered structure that best corresponds.

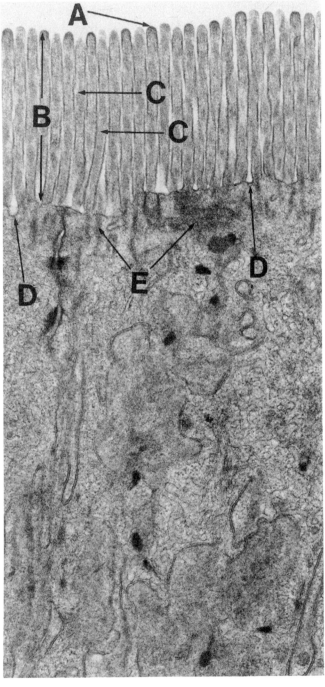

Figure 3.8

259. The projections indicated at letter B are

(A) cilia
(B) microvilli
(C) flagella
(D) basal bodies
(E) focal adhesion points

260. The structure indicated at the letter A is

(A) the basement membrane
(B) an endocytic vesicle
(C) a microtubule cut in longitudinal section
(D) the plasma membrane
(E) part of the Golgi apparatus

261. Which protein is most likely to be found in the linear structures labelled C?

(A) histone H-1
(B) collagen
(C) actin
(D) desmin
(E) clathrin

262. Material at letter E is part of the

(A) extracellular matrix
(B) terminal web
(C) smooth endoplasmic reticulum
(D) phagolysosome
(E) basal lamina

263. At the letters D, which process is most likely to occur?

(A) fatty acid synthesis
(B) steroid secretion
(C) photosynthesis
(D) receptor-mediated endocytosis
(E) nucleation of microtubule assembly

264. The spacing between adjacent structures labelled B is closest to which of the following values?

 (A)   1.0 μm
 (B)   0.1 μm
 (C)   10 nm
 (D)   10 μm
 (E)   5 μm

ANSWERS AND TUTORIAL ON ITEMS 259-264

The answers are: **259-B; 260-D; 261-C; 262-B; 263-D; 264-B. Figure 3.8** is a transmission electron micrograph of the brush border of intestinal epithelial absorptive cells. The brush border consists of numerous closely packed cellular projections called **microvilli** (B). Microvilli are about 0.1 μm in diameter and 0.5 to 1.0 μm long; in the brush border they are tightly packed together so that the spacing between adjacent microvilli is about the same as the diameter of each individual microvillus. Each microvillus is surrounded by a **plasma membrane** (A) that is part of the plasma membrane of the epithelial cell. Each microvillus contains a core bundle of **microfilaments** (C) made of the protein **actin**. These microfilaments project into the terminal web (E), a structure that lends support to the microvilli. The microvilli serve to provide an extended surface of plasma membrane, providing the space for the transport channels and enzymes needed so these absorptive cells can carry out their principal function. These epithelia are also involved in the active uptake of small macromolecules (peptides, fragments of nucleic acids, etc.) via **receptor mediated endocytosis**. This involves the formation of numerous **endosomes**, a process that occurs as the bases of the **clefts** between adjacent microvilli (D).

CHAPTER IV
STRUCTURE AND FUNCTION OF CYTOPLASMIC ORGANELLES

Items 265-276

Most human cells are composed of a complex collection of organelles. Match the organelle in the answers below with the most appropriate description of its physiologic function in the items below.

(A)   Plasma membrane
(B)   Nucleus
(C)   Mitochondria
(D)   Golgi apparatus
(E)   Rough endoplasmic reticulum
(F)   Smooth endoplasmic reticulum
(G)   Microtubule
(H)   Microfilament
(I)   Intermediate filament
(J)   Glycogen granule
(K)   Pigment granule
(L)   Lipid droplet
(M)   Lysosome
(N)   Peroxisome

265.   A large organelle surrounded by two membranes, forming an envelope interrupted by many pores with octagonal symmetry.

266.   Consists of flattened stacks of membrane-delimited cisternae studded with ribosomes active in the synthesis of proteins destined for secretion.

267.   This cytoskeletal element functions in certain forms of cell movement and is an essential component of the mitotic spindle and centrioles.

268.   A cytoskeletal element comprised of the protein actin.

269.   This structure contains a tyrosine-derived polymer that absorbs UV light and functions to reduce the incidence of mutagenesis in epidermal cells.

270. Abundant in leucocytes, this organelle contains a complex array of hydrolytic enzymes that function in the degradation of "worn out" intracellular organelles.

271. Another membranous organelle surrounded by two membranes, this contains proteins critical for electron transport and synthesis of compounds with high energy phosphate bonds.

272. A structure that serves to store cholesterol and other precursors of steroid hormones.

273. A cytoskeletal element that may contain desmin or vimentin.

274. In skeletal and cardiac muscle cells, this organelle sequesters large stores of Ca^{2+} which are released into the cytoplasm subsequent to depolarization of the plasma membrane.

275. This organelle is surrounded by a single membrane, contains high concentrations of the enzyme catalase, and carries out the breakdown of long chain fatty acids.

276. This organelle requires a low internal pH for the function of many of its proteins.

ANSWERS AND TUTORIAL ON ITEMS 265-276

The answers are: **265-B; 266-E; 267-G; 268-H; 269-K; 270-M; 271-C; 272-L; 273-I; 274-F; 275-N; 276-M.** The **plasma membrane** (A) is a **phospholipoprotein bilayer** (unit membrane) surrounding all cells in the human body. It serves as the functional boundary between a cell and its environment. Its selective permeability characteristics regulate cell volume, nerve conduction and muscle contraction. It is also the site for many hormone receptors and other macromolecules involved in cell-cell communication and adhesion.

The **nuclear envelope** encloses the **nucleus** (B). Its intermembranous space is continuous with the lumen of the endoplasmic reticulum. The outer surface of the nuclear envelope binds ribosomes. The inner surface is often covered by clumps of heterochromatin. The nuclear envelope is perforated by large octagonal pores that allow the passage of mRNA which is synthesized in the nucleus but transported to the cytoplasm where it is translated into proteins on ribosomes.

Mitochondria (C) are intracellular organelles consisting of two unit membranes. The outer membrane is relatively permeable and thus allows substrates to enter mitochondria. The inner membrane, which is relatively impermeable, is highly folded into cristae or tubular projections into the mitochondrial matrix. Within the matrix and on inner membranes, reduced nucleotides produced by oxidation of substrates are converted into ATP via the electron transport chain.

The **Golgi apparatus** (D) consists of a collection of flattened membranous cisternae and membrane-delimited vesicles. It is involved in glycosylation and packaging of many proteins synthesized on the ribosomes of the **rough endoplasmic reticulum** (E) and destined for secretion from cells. The Golgi apparatus is also involved in the production of **lysosomes** (M). These membrane-bound organelles contain hydrolytic enzymes used to degrade engulfed materials such as bacteria and worn-out intracellular organelles.

The **smooth endoplasmic reticulum** (F) is a membranous organelle consisting of an anastomosing network of interconnected cisternae and tubules. It functions in glycogen breakdown, synthesis of cholesterol and phospholipids and serves to detoxify drugs and poisons. In muscle cells, the sarcoplasmic reticulum, a variety of smooth endoplasmic reticulum, sequesters Ca^{2+} and is important for regulating free Ca^{2+} around myofibrils.

Within every cell, there is a supporting framework of tubules and filaments known as the **cytoskeleton**. It consists of three distinct intracellular nonmembranous organelles: (1) **microtubules**, (2) **microfilaments**, and (3) **intermediate filaments**. **Microtubules** (G) are important elements of the mitotic spindle. They attach to chromosomes at the kinetochore and are essential for mitotic separation of chromosomes. Microtubules are also important constituents of centrioles, basal bodies, cilia and flagella. Therefore, they are involved in movement of some cells, e.g., spermatozoa. **Microfilaments** (H) are actin-rich filamentous structures. In nonmuscle cells, they provide a dynamic framework for the cell permitting extensions of pseudopodia, endocytosis of extracellular materials, and cell motility. They are especially abundant in muscle cells where they are essential component of the thin filaments. **Intermediate filaments** (I) are fibrous structures consisting, depending on location, of several different proteins, including desmin, vimentin, and keratin.

Glycogen granules (J) and **pigment granules** (K) are examples of intracellular inclusions. Glycogen is a stored glucose polymer than can be degraded to release glucose when the individual is hypoglycemic. Pigment granules contain a tyrosine-rich polymer called melanin. The aromatic side groups of the constituent amino acids absorb UV radiation from sunlight and thus shield nuclear DNA from mutation. This is why there are many melanin granules forming an umbrella-shaped cap over the nuclei of cells in the epidermal stratum basale.

Lipid droplets (L) are abundant in steroid synthesizing cells such as those found in the adrenal cortex or interstitial (Leydig) cells of the testes. Here they function as lipid precursors for steroid hormones. In addition, lipid droplets are found in adipocytes in fat. Here they serve as a storage form of triglycerides which can be utilized in generation of energy when food intake is less than the energy output.

Lysosomes (M) are membrane-bound organelles containing hydrolytic enzymes. They are produced in the Golgi apparatus. They have several functions in cellular physiology, including degradation of phagocytosed foreign materials (e.g., bacteria), degradation of lipid aggregates and glycogen granules (e.g., in glycogen storage diseases such as Pompe's disease, lysosomal hydrolases are congenitally absent and glycogen accumulates in cells), and tissue degradation during regression (e.g., following pregnancy, uterine regression aided by lysosomes). Most of the enzymes found in lysosomes have slightly acid pH optima and the lysosomal interior is maintained at a pH lower than the cytoplasm. Lysosomes are abundant in macrophages and in a highly modified form in granulocytes as specific granules.

Peroxisomes (N) are organelles surrounded by a single membrane. They contain a number of enzymes that carry out breakdown reactions in which hydrogen peroxide is generated. They also contain enzymes, such as catalase, that degrade peroxides. Peroxisomes are responsible for the breakdown of long chain fatty acids.

A one month-old male infant became progressively weak and listless. He developed enlargement of the heart and abnormalities on the EKG with enlarged QRS complexes and a shortened PR interval. Biopsy of skeletal muscle followed by electron microscopy revealed moderately electron dense granular deposits in membrane delimited vesicles. The child died of heart failure in his 18th month.

277.   The most likely diagnosis of this disease is

(A)   Sandhoff disease
(B)   von Gierke's disease
(C)   cystic fibrosis
(D)   Pompe's disease
(E)   Duchenne muscular dystrophy

278.   The organelle most effected in this disease is the

(A)   nuclear envelope
(B)   plasma membrane
(C)   mitochondria
(D)   lysosome
(E)   microtubule

279.   Which enzyme would be lacking in this disease?

(A) α-1,6 glucosidase
(B) β-glucuronidase
(C)   neuraminidase
(D) β-galactosidase
(E) α-1,4 glucosidase

280.   The granular material seen in electron microscopic biopsies is

(A)   glycolipids
(B)   lipofuscin
(C)   amyloid deposits
(D)   glycogen
(E)   proteoglycan aggregates

ANSWERS AND TUTORIAL ON ITEMS 277-280

The answers are: **277-D; 278-D; 279-E; 280-D**. This child suffers from **Pompe's disease (Glycogenosis, Type II A)**. This is a rare (1/400,000 live births) fatal disease with no known cure. Most children show onset of symptoms in the first two months of life and rarely live beyond two years. Pompe's disease is caused by an **autosomal recessive mutation** causing absence or severe reduction in the amount of the lysosomal enzyme **α-1,4-glucosidase**. Under normal conditions, this enzyme is responsible for degradation of **glycogen** which is destined to be used for energy production. When glycogen can not be degraded, it accumulates inside membrane bound vesicles that are derived from lysosomes. Massive deposits of glycogen eventually compromise muscular function, leading to weakness, cardiac myopathy and death, usually from heart failure.

Items 281-287

The following is a list of mitochondrial enzymes. Match the most appropriate description of the enzyme in the numbered items with the lettered enzyme in the list below:

- (A) Monoamine oxidase
- (B) Phospholipase A
- (C) Kynurenine hydroxylase
- (D) Adenyl kinase
- (E) Succinate dehydrogenase
- (F) NADH dehydrogenase
- (G) Cytochrome c oxidase
- (H) Pyruvate dehydrogenase
- (I) Isocitrate dehydrogenase
- (J) Ornithine transcarbamoylase

281. This enzyme is present in the outer mitochondrial membrane. It is part of the pathway for L-tryptophan catabolism.

282. This important TCA cycle enzyme is found in the mitochondrial matrix.

283. This enzyme is present in the outer mitochondrial membrane. It inactivates epinephrine.

284. This enzyme is found in the intermembrane space of mitochondria.

285. This important constituent of the electron transport chain is found in the inner mitochondrial membrane. It is a FMN-linked dehydrogenase.

286. This important constituent of the electron transport chain is found in the inner mitochondrial membrane. It catalyzes the terminal step in the electron transport chain.

287. This enzyme is found in the mitochondrial matrix. It is a massive multi-subunit complex with a MW = 8 x 10^6.

ANSWERS AND TUTORIAL ON ITEMS 281-287

The answers are: **281-C; 282-I; 283-A; 284-D; 285-F; 286-G; 287-H**. **Mitochondria** are complex cellular organelles consisting of an **outer membrane**, an **intermembrane space**, an **inner membrane** with numerous folds called **cristae** and an **internal amorphous matrix**. There is some evidence that mitochondria have evolved from endosymbiotic organisms. Mitochondria are the powerhouses of the cell. They have the enzyme systems for the oxidation of pyruvate and fatty acids, TCA cycle enzymes, and systems for oxidative phosphorylation and electron transport. The most important metabolic product of mitochondria is the energy rich compound **ATP**. There is a tremendous variation in the number and structure of mitochondria in different cell types. For example, erythrocytes have no mitochondria at all. Liver parenchymal cells have 1,000 or more mitochondria per cell, reflecting their complex catabolic functions. Cardiac muscle cells have many mitochondria producing the ATP required for continuous cardiac muscle contraction.

The **outer mitochondrial membrane** is relatively simple on a structural and functional level. It consists of approximately 50% lipid and 50% protein and has relatively few transport or enzymatic functions. The enzymes **monoamine oxidase** (A), **phospholipase A** (B) and **kynurenine hydroxylase** (C) are all found in the outer mitochondrial membrane. The **intermembrane space** lies between the outer and inner mitochondrial membranes. **Adenyl kinase** (D) and **nucleoside diphosphate kinase** have been found there.

The **inner mitochondrial membrane** is much more complex than the outer mitochondrial membrane and contains approximately 75% protein. Here, one finds **succinate dehydrogenase** (E); all three components of the **electron transport chain**: NADH dehydrogenase (F), and cytochrome c oxidase (G); and **enzymes for ATP synthesis** (ATP synthase) and ADP-ATP transport (ADP-ATP translocase). In addition, the inner mitochondrial membranes have enzymes for transport of carboxylates, glutamate-aspartate and Ca^{2+}.

The **mitochondrial matrix** contains pyruvate dehydrogenase (H), citrate synthase, isocitrate dehydrogenase (I), all of the other enzymes of the **TCA cycle** except for succinate dehydrogenase (E), succinyl CoA synthetase, fatty acid β-oxidation enzymes, glutamate dehydrogenase, glutamate-oxaloacetate transaminase and ornithine transcarbamoylase (J).

Items 288-290

The 8 year-old daughter of Jewish parents whose grandparents were born in Russia has hepatosplenomegaly and painful bone lesions. Urinalysis reveals high levels of the neutral sphingolipid, 1-β-glucoceramide.

288.   The most likely diagnosis is

   (A)   Tay-Sachs disease
   (B)   Gaucher's disease
   (C)   Fabry's disease
   (D)   Niemann-Pick disease
   (E)   Hurler syndrome

289.   The intracellular organelle most effected by this disease is the

   (A)   mitochondrion
   (B)   centriole
   (C)   lysosome
   (D)   rough endoplasmic reticulum
   (E)   Golgi apparatus

290.   Which enzyme is altered in this disease?

   (A)   glucocerebrosidase
   (B)   glucuronidase
   (C)   gangliosidase
   (D)   galactosyltransferase
   (E)   chymotrypsin

ANSWERS AND TUTORIAL ON ITEMS 288-290

The answers are: **288-B; 289-C; 290-A**. This child suffers from **Gaucher's disease**. There are many different mutations including point mutations, deletions and crossover events that lead to this disease. Gaucher's disease is most common in Ashkenazi Jews. The secretion of **glucocerebroside**, a neutral sphingolipid formed by the covalent coupling of glucose to ceramide, is characteristic of Gaucher's disease. This disease is due to congenital abnormalities in the function of a **lysosomal enzyme** for degradation of glucocerebroside, **glucocerebrosidase**. Lysosomes are used to degrade endogenous molecular components of cells. They are particularly abundant in macrophages. Although the mutant genotype is present in all cells of the body, it is

109

most often expressed only in macrophages, leading to hepatosplenomegaly and erosion of long bones.

Tay-Sachs disease, caused by a deficiency in **hexosaminidase A**, leads to the accumulation of **ganglioside G_{M2}**. **Fabry's disease**, caused by a deficiency in α-**galactosidase A**, leads to the accumulation of **ceramide trihexoside**. **Niemann-Pick disease**, caused by a deficiency in **sphingomyelinase**, leads to the accumulation of **sphingomyelin**. **Hurler syndrome**, caused by a deficiency in α-**L-iduronidase**, leads to the accumulation of **dermatan sulfate** and **heparan sulfate**.

Items 291-294

A 2 year-old first child, a female, presents with blindness, a red spot in the macula lutea and mental retardation. The parents both have a Jewish ethnic background. They first noticed a problem when the child was 6 months old, indicating that she didn't seem to look at their faces and showed attention deficits. Biochemical tests indicate that the child is suffering from Tay-Sachs disease.

291.   This disease results in the accumulation of which of the following gangliosides?

   (A)   ganglioside G_{M2}
   (B)   ganglioside G_{M1}
   (C)   glucocerebroside
   (D)   sphingomyelin
   (E)   galactocerebroside

292.   Which enzyme is defective in Tay-Sachs disease?

   (A)   galactocerebrosidase
   (B)   glucocerebrosidase
   (C)   α-Galactosidase A
   (D)   arylsulfatase A
   (E)   hexosaminidase A

293.   The defective enzymes are located in which organelle?

   (A)   nuclear envelope
   (B)   rough endoplasmic reticulum
   (C)   lysosome
   (D)   Golgi apparatus
   (E)   mitochondria

294. The risk of having a second afflicted child when a first child has been afflicted is approximately

   (A)   1/100,000
   (B)   1/1,000
   (C)   1/100
   (D)   1/10
   (E)   1/4

ANSWERS AND TUTORIAL ON ITEMS 291-294

The answers are: **291-A; 292-E; 293-C; 294-E**. There is a group of **lysosomal storage diseases** involving abnormalities in the catabolism of sphingolipids. Ganglioside G_{M1} (ceramide-glc-gal-NANA-galNAC-gal) is degraded to G_{M2} (ceramide-glc-gal-NANA-galNAC) by β-galactosidase. A deficiency in this step leads to generalized gangliosidosis and the accumulation of G_{M1}. G_{M2} is degraded to ceramide-glc-gal-NANA by β-hexosaminidase A. A deficiency in this step leads to **Tay-Sachs disease** and the accumulation of G_{M2}. Ceramide-glc-gal-NANA is degraded to ceramide-glc-gal by neuraminidase. Ceramide-glc-gal is degraded to ceramide-glc (glucocerebroside) by β-galactosidase. Ceramide-glc is degraded to ceramide by β-glucosidase. A deficiency in this enzyme leads to **Gaucher's disease** and accumulation of glucocerebroside.

Tay-Sachs disease (G_{M2}-gangliosidosis with hexosaminidase A deficiency) is a lysosomal storage disease caused by an autosomal recessive mutation. Tay-Sachs disease has a theoretical incidence of 1/3,800 live births among Ashkenazi Jews. It is also fairly common among French-Canadians. The parents are heterozygous carriers with reduced levels of enzymes but without manifestations of the disease. Parental carriers can be detected by enzyme assays in cultured cells. The incidence can be reduced by carrier screening and prenatal diagnosis. When both parents are heterozygotes, which must be true for the child to be afflicted, then the chance of any of their children being afflicted is 1/4. These parents will have 1/4 homozygous wild type (normal), 2/4 heterozygotes (carrier), and 1/4 homozygous recessive (diseased). Homozygous recessive individuals usually die between 2 and 4 years of age; they never reach reproductive age.

Items 295-298

A 4 year-old female suffers from skeletal deformities, corneal opacity, mental retardation and excretes dermatan sulfate and heparan sulfate in her urine. Both parents appear normal. Biochemical analysis of skin fibroblasts isolated from the child show severe reductions in an enzyme for glycosaminoglycan degradation. Analysis of fibroblasts from both parents show moderate depression of the level of the same enzyme. Further tests lead to a diagnosis of mucopolysaccharidosis I-H (Hurler syndrome).

295.   The most likely defective organelle in this disease is the

   (A)   Golgi apparatus
   (B)   endoplasmic reticulum
   (C)   peroxisome
   (D)   lysosome
   (E)   mitochondrion

296.   This type of mucopolysaccharidosis is caused by a reduced level of the enzyme

   (A)   β-D-glucuronidase
   (B)   α-L-iduronidase
   (C)   hyaluronidase
   (D)   β-D-galactosidase
   (E)   N-acetyl-β-D-galactosidase

297.   The most abundant urinary glycosaminoglycan(s) in this disease is (are)

   (A)   chondroitin sulfate and hyaluronic acid
   (B)   dermatan sulfate and hyaluronic acid
   (C)   hyaluronic acid
   (D)   heparin
   (E)   dermatan sulfate and heparan sulfate

298.   The excreted mucopolysaccharides both contain which monosaccharides?

   (A)   iduronic acid and amino sugars
   (B)   glucuronic acid and galactose
   (C)   glucuronic acid and glucose
   (D)   N-acetylgalactosamine and galactose
   (E)   N-acetylglucosamine and glucose

ANSWERS AND TUTORIAL ON ITEMS 295-298

The answers are: **295-D; 296-B; 297-E; 298-A**. This child suffers from mucopolysaccharidosis I-H (**Hurler syndrome**), a disease caused by an **autosomal recessive mutation**. The coarse facial features, hepatosplenomegaly, corneal opacity and skeletal deformities are suggestive of Hurler syndrome. This disease is caused by a deficiency in the lysosomal enzyme α-**L-iduronidase** which is normally used in the degradation of **dermatan sulfate** and **heparan sulfate**. These glycosaminoglycans both contain L-iduronic acid and D-glucuronic acid. Dermatan sulfate contains N-acetyl-galactosamine while heparan sulfate contains N-acetyl-glucosamine. When α-L-iduronidase is lacking, dermatan sulfate and heparan sulfate are excreted in large amounts in the urine and accumulate in the tissues of the body. Deficiencies in mucopolysaccharide turnover often result in developmental defects in the musculoskeletal system because cartilage and bone formation require normal mucopolysaccharide synthesis and degradation. Accumulation of mucopolysaccharides in leucocytes and hepatic parenchymal cells usually leads to hepatosplenomegaly.

Items 299-305 refer to the electron micrograph (**Figure 4.1**) of a plasma cell. For each numbered item, select the most appropriate lettered organelle or region.

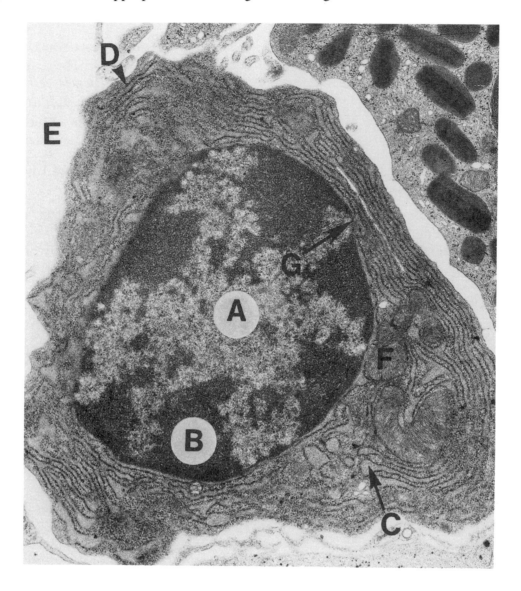

Figure 4.1

299. Messenger RNA (mRNA) is synthesized here.

300. Once synthesized, mRNA enters the cytoplasm here.

301. DNA that is not being transcribed to mRNA is most likely to be found in this location.

302. The completed light and heavy chains of synthesized immunoglobulins are first found here.

303. Once secreted from cells, immunoglobulins enter this space.

304. Peptidyl bond formation is carried out here.

305. Energy required for coupling of amino acids to transfer RNAs (tRNAs) is derived from hydrolysis of a compound synthesized in this organelle.

ANSWERS AND TUTORIAL ON ITEMS 299-305

The answers are: **299-A; 300-G; 301-B; 302-C; 303-E; 304-D; 305-F. Figure 4.1** is a plasma cell. **Plasma cells** synthesize **immunoglobulins** by mechanisms common to most cells engaged in production of proteins for export from the cell. Information regarding the amino acid sequence of the secretory protein is stored, as chromatin, in the nucleus. That information is transcribed into the codon sequence of messenger RNA (mRNA) synthesized in regions of the nucleus where the chromatin is decondensed and present as **euchromatin** (A). DNA sequences that are NOT being transcribed are maintained in the condensed state of chromatin - called **heterochromatin** (B). Once an mRNA has been synthesized in the nucleus it reaches the cytoplasm by passing through gaps - called **nuclear pores** (G) - in the nuclear envelope. Amino acids that will be coupled together during protein synthesis are "activated" by combination with specialized carrier RNAs, called tRNAs; the energy required for this activation is derived from the hydrolysis of ATP, made in **mitochondria** (F). The information encoded in an mRNA is translated into an amino acid sequence by ribosomes, which "read" the messenger RNA information and perform the essential step of coupling activated amino acids by forming peptidyl bonds. Secretory proteins are synthesized by ribosomes attached to the membranes of the **rough endoplasmic reticulum** (rER, D). As the nascent polypeptide chain emerges from the large subunit of the bound ribosome, it is transferred through the membrane and into the **lumen of the rER** (C), where it is released from the ribosome. The completed polypeptide chain is subjected to a number of processing steps, many of these (such as addition and removal of specific carbohydrates) occur in the Golgi apparatus (not shown).

A despondent lover from New Iberia, LA, nicknamed "Cajunman", became severely depressed after his fiance terminated their relationship. He attempted suicide by ingesting a large dose of seconal. Fortunately, he called a friend soon after ingesting the drug. The friend rushed him to the emergency room where treatment was administered.

306. Which organ in his body is most important for seconal detoxification?

   (A)   thyroid
   (B)   liver
   (C)   pancreas
   (D)   stomach
   (E)   small intestines

307. Which organelle in the parenchymal cells of the most important organ above would be most altered by a large dose of seconal?

   (A)   rough endoplasmic reticulum
   (B)   smooth endoplasmic reticulum
   (C)   Golgi apparatus
   (D)   nucleus
   (E)   microtubules

308. Which enzyme system is most crucial for detoxification of seconal?

   (A)   mitochondrial cytochrome oxidase
   (B)   microsomal NADPH-dependent cytochrome P_{450}
   (C)   glycosyltransferase
   (D)   glucuronyltransferase
   (E) β-galactosidase

309. Seconal is detoxified initially by which reaction?

   (A)   hydroxylation
   (B)   decarboxylation
   (C)   conjugation
   (D)   deamination
   (E)   transamination

The answers are: **306-B; 307-B; 308-B; 309-A**. **Seconal** is a **barbiturate**. It is used as a sedative but can be lethal in high doses due to respiratory depression. The liver is the crucial organ in the body for detoxification of ingested poisons. In the case of barbiturate poisoning, liver parenchymal cells show a rapid increase in the amount of **smooth endoplasmic reticulum**. These membranes have an enzyme system for hydroxylation of barbiturates, thus making them more water-soluble and therefore more easily eliminated from the body. This detoxification reaction is catalyzed by the **NADPH-dependent cytochrome P_{450}** of the smooth endoplasmic reticulum. When cells are homogenized and subjected to differential centrifugation, the smooth endoplasmic reticulum fragments end up in the **microsomal fraction**.

Items 310-314

310.   All of the following are a function of hepatocytes **EXCEPT**:

   (A)   conversion of ammonia to urea
   (B)   synthesis of circulating lipoproteins
   (C)   production of bile
   (D)   secretion of immunoglobulins
   (E)   storage of glycogen

311.   *Amanita phalloides* produces phalloidin which binds to and hyperstabilizes actin in the polymeric form. Phalloidin damages the liver, eventually leading to death. At earlier stages of phalloidin toxicosis, which of the following is **MOST LIKELY** found in electron microscopic studies of a liver biopsy?

   (A)   disruption of desmosomal contacts between hepatocytes
   (B)   numerous cells arrested in metaphase of mitosis
   (C)   thickening of arrays of cortical microfilaments near bile canaliculi
   (D)   presence of Ca^{2+} precipitates within the smooth endoplasmic reticulum
   (E)   depletion of crystalline deposits in peroxisomes

312.   Liver failure due to phalloidin-poisoning may cause coma because

   (A)   elevated levels of blood ammonia are toxic to the central nervous system
   (B)   portal blood flow is increased
   (C)   insulin secretion from hepatocytes is blocked
   (D)   protein synthesis by astrocytes is blocked due to increased levels of G-actin
   (E)   amino acid catabolism decreases

313. Bilirubin

    (A)   is a water-soluble compound
    (B)   produced by Kupffer cells is taken up by hepatocytes
    (C)   derives primarily from the breakdown of lymphocytes
    (D)   is normally secreted by hepatocytes into the blood stream
    (E)   conjugation to glucuronic acid occurs in secondary lysosomes

314. Jaundice results from all of the following **EXCEPT**:

    (A)   slow development of the smooth endoplasmic reticulum in newborns
    (B)   blockage of the cystic duct
    (C)   destruction of tight junctions around bile canaliculi
    (D)   enterohepatic circulation of bile salts
    (E)   abnormally-high rates of degradation of erythrocytes

ANSWERS AND TUTORIAL ON ITEMS 310-314

The answers are: **310-D; 311-C; 312-A; 313-B; 314-D**. The liver produces many of the proteins found in the bloodstream. An exception is the production of circulating antibodies, which is the function of plasma cells derived from B-lymphocytes. One of the principal functions of hepatocytes is the production and secretion of **bile**. Bile contains a mixture of compounds, including cholesterol, phospholipids and bile acids. **Kupffer cells** (mononuclear phagocytes in the liver) produce bilirubin from the hemoglobin released during the breakdown of senescent erythrocytes. The **bilirubin**, which is insoluble in water, is taken up by hepatocytes; and, in the smooth endoplasmic reticulum, is conjugated to glucuronic acid, producing bilirubin glucuronide, a water soluble bile acid. Bile acids and the other components of bile are released from hepatocytes into the lumen of **bile canaliculi**. **Microfilaments** (polymeric forms of actin and associated cytoskeletal proteins) are a principal component of the cortical cytoplasm around bile canaliculi and are presumed to regulate bile secretion. The phalloidin-induced increase in the microfilament meshwork around bile canaliculi is thought to block normal bile release and cause bile to back up into vascular circulation, causing **jaundice**. Jaundice (excessive levels of circulating bilirubin) results from blockage of bile release from hepatocytes as well as from blockage of the ducts that carry bile from the liver to its storage point (gallbladder) and its place of usage (lumen of the intestines).

Newborns often have underdeveloped hepatocyte smooth endoplasmic reticulum, which leads to release of bilirubin into the bloodstream and thus jaundice. Most of the bile acids that reach the lumen of the small intestine are (normally) reabsorbed by the intestinal mucosa and are returned to the liver (enterohepatic circulation) to be reused.

Another function of the liver is to metabolize amino acids. The ammonia generated during amino acid breakdown is converted by the liver to urea, which is excreted in urine. In terminal stages of liver failure, conversion of ammonia to urea is not rapid enough to prevent elevated levels of ammonia in the blood. Coma often arises because ammonia is toxic to the CNS.

315. Which are sites for protein synthesis in mammalian cells?

(A)   ribosomes attached to membranes of the rough endoplasmic reticulum
(B)   mitochondrial ribosomes
(C)   ribosomes free in the cytoplasm
(D)   All of the above
(E)   None of the above

316. All of the following statements about ribosomes are correct **EXCEPT**:

(A)   All ribosomes contain both a large subunit and a small subunit.
(B)   Polysomes consist of a single mRNA associated with many ribosomes.
(C)   Except for the 5S RNA, ribosomal RNAs are synthesized in the nucleolus.
(D)   Ribosomes synthesize the proteins from which ribosomes are made.
(E)   Transfer RNAs are made on single ribosomes.

317. All of the following inhibit protein synthesis **directly EXCEPT**:

(A)   puromycin
(B)   cycloheximide
(C)   chloramphenicol
(D)   diphtheria toxin
(E)   actinomycin

318. All of the following contain RNA **EXCEPT**:

(A)   tRNA
(B)   large ribosomal subunit
(C)   small ribosomal subunit
(D)   polysome
(E)   aminoacyl-tRNA synthetase

319. Sickle cell disease

(A)   reflects a failure to produce erythrocytes
(B)   is a defect in the binding of mRNA to ribosomes
(C)   is caused by a mutation that alters the amino acid sequence of the β-chain of hemoglobin
(D)   can be cured by exposing patients to reduced levels of oxygen
(E)   leads to the production of an abnormally long α-chain of hemoglobin

320. All of the following statements regarding protein synthesis in mammalian cells are true **EXCEPT**:

    (A)   The amino acid sequence of a protein is determined by codons, with each codon being a sequence of four bases in the messenger RNA.

    (B)   Codon information in DNA is transcribed into mRNA and the information in the mRNA is translated into the amino acid sequence of the protein.

    (C)   Initiation of protein synthesis requires several initiation factors, which are transiently associated with the ribosome.

    (D)   Ribosomes read the codon information in mRNA from the 5' to the 3' end.

    (E)   Proteins are synthesized starting from the amino terminus and proceeding toward the carboxyl terminus.

ANSWERS AND TUTORIAL ON ITEMS 315-320

The answers are: **315-D; 316-E; 317-E; 318-E; 319-C; 320-A**. With a few exceptions, protein synthesis in mammalian cells occurs on **ribosomes**. The amino acid sequence of a protein is determined by sequence information contained as a series of **codons** (triplets of bases) in the DNA. This information is **transcribed** into a corresponding sequence of base triplets in **messenger RNA** (mRNA). Each codon corresponds to one amino acid in the final synthesized protein. Ribosomes are made in the **nucleolus**; the assembly process involves the production of all rRNAs (the 5S RNA is made in the nucleus, but **not** in the nucleolus) and their association with the various ribosomal proteins, all of which are made in the cytoplasm on pre-existing ribosomes.

    All ribosomes have a **large subunit** and a **small subunit**, each with distinct rRNAs and proteins. It is rare for ribosomes to function as single units (monosomes); in most cases multiple ribosomes associate with the same mRNA to form a **polysome**, with all of the constituent ribosomes reading the same RNA in sequence. Ribosomes "read" the codon information in mRNA, proceeding from the 5' to the 3' end of the mRNA. This corresponds to synthesis beginning at the amino terminus of the protein and proceeding to the carboxyl terminus. Alteration of a single base in a codon can lead to an altered protein whose function may be altered or stopped completely. **Sickle cell anemia** is caused by a single base alteration in the gene coding for the β-chain of hemoglobin. Individuals homozygous for this altered gene have erythrocytes whose hemoglobin, when deoxygenated, forms polymeric arrays that distort the erythrocyte, rendering it inflexible and leading to a shortened life span for the cell and reducing the ability of the cell to pass through capillaries such as those in the lung.

    In addition to mRNA, ribonucleic acids are involved in other ways in protein synthesis. Both the small and large subunits of ribosomes contain specific (ribosomal) RNA species. The mechanism by which ribosomes "read" the codon information in mRNA involves a set of small RNA molecules (transfer or tRNAs) that are coupled to specific amino acids by enzymes called aminoacyl-tRNA synthetases. tRNA molecules are adaptors that have codon-specific sites by which they interact with messenger RNA. Thus specific proteins (aminoacyl-tRNA synthetases) are critical in assuring that information in nucleic acid sequences are converted to information in the amino acid sequences of proteins. The process of protein synthesis is initiated by a number of factors that associate transiently with the ribosomal subunits. Initiation is followed by elongation and then termination, stages that also involve factors transiently binding to ribosomes and/or mRNA.

In addition to ribosomes free in the cytoplasm and bound to membranes of the **rough endoplasmic reticulum** (rER), mammalian cells also contain ribosomes in the **mitochondria**. All three classes of ribosomes are active in protein synthesis. Mitochondrial ribosomes are thought to be evolutionarily related to procaryotic (e.g., bacterial) ribosomes. Mitochondrial protein synthesis is inhibited by the antibiotic **chloramphenicol**, which acts on the large subunit of procaryotic ribosomes. Other compounds that block protein synthesis directly include **cycloheximide** (acts on eukaryotic large subunits), **puromycin** (a tRNA analog that causes premature termination) and **diphtheria toxin** (inactivates one of the elongation factors). Protein synthesis can also be blocked indirectly, as for example by actinomycin, which blocks the synthesis of mRNA.

Items 321-329

321. All of the following completions are true **EXCEPT**:

Ribosomes bound to the membranes of the rough endoplasmic reticulum

(A) synthesize proteins that are to be secreted by the cell.
(B) contain ribosomal RNAs that are different from those in free ribosomes.
(C) produce lysosomal enzymes.
(D) are attached to the rough endoplasmic reticulum via transient connections between the membrane and the large ribosomal subunit.
(E) make proteins that become integral components of the endoplasmic reticulum membranes.

322. The signal sequence that directs the synthesis of secretory proteins by membrane-bound ribosomes

(A) is at the 3′ end of the mRNA, immediately preceding the termination codon
(B) forms part of the large ribosomal subunit RNA
(C) codes for a sequence of hydrophobic amino acids at the amino terminus of the nascent protein
(D) normally is cleaved from the mRNA soon after initiation has been completed
(E) must be recognized by a receptor in the Golgi apparatus if proper polypeptide synthesis is to proceed

323. Synthesis of lysosomal proteins **and** transfer of those proteins to the lumen of the rER requires all of the following **EXCEPT**:

    (A) a signal sequence on the mRNA
    (B) ribophorins in the rER membranes
    (C) signal recognition particle function
    (D) docking protein in the rER membranes
    (E) signal peptidase on the ribosomes

324. All of the following proteins are synthesized by ribosomes bound to the rER **EXCEPT**:

    (A) signal peptidase
    (B) ribophorin I
    (C) peptidyl transferase
    (D) collagen
    (E) serum albumin

325. Glycosylation of secretory proteins

    (A) begins when the nascent polypeptide is in the lumen of the rER
    (B) occurs at amino acid residues coded for by the signal sequence
    (C) requires the presence of mannose 6-phosphate
    (D) only occurs in the alkaline environment of condensing vacuoles
    (E) involves a number of hydrolytic enzymes found in lysosomes

326. Each of the following is a **co**-translational modification of a protein destined to be secreted from the cell of origin **EXCEPT**:

    (A) proteolytic cleavage by signal peptidase
    (B) rearrangement of disulfide bonds
    (C) addition of core oligosaccharides from dolichol
    (D) addition of O-linked oligosaccharides
    (E) hydroxylation of proline residues

327. The Golgi apparatus carries out all of the following **EXCEPT**:

    (A) addition of sugars to the oligosaccharide core
    (B) synthesis of mannose 6-phosphate receptors
    (C) removal of sugars from the oligosaccharide core
    (D) addition of mannose 6-phosphate to lysosomal enzymes
    (E) sorting of lysosomal enzymes from proteins destined for secretion

328. Enzymes required for formation and function of peroxisomes are synthesized

    (A)   on ribosomes attached to the rER
    (B)   on free cytoplasmic ribosomes
    (C)   on ribosomes bound to the peroxisome membrane
    (D)   in the matrix of the peroxisomes
    (E)   by mitochondrial ribosomes

329. Proteins that are destined for insertion into the mitochondrial membrane and into peroxisomes are carried to those locations in association with

    (A)   lipid droplets
    (B)   ribosomal precursor complexes
    (C)   molecular chaperones
    (D)   the signal recognition particle
    (E)   transfer RNA

ANSWERS AND TUTORIAL ON ITEMS 321-329

The answers are: **321-B; 322-C; 323-E; 324-C; 325-A; 326-D; 327-B; 328-B; 329-C**. Ribosomes bound to the membranes of the rough endoplasmic reticulum (rER) synthesize a set of proteins different from the set synthesized on free cytoplasmic ribosomes. The free and bound ribosomes are identical in terms of protein and rRNA composition. The differences in their synthetic activities reflect a difference in the mRNAs that they translate. Ribosomes attached to the rER read mRNAs that have a **signal sequence of codons** near the 5′ end. This signal sequence codes for a set of 20-25 hydrophobic amino acids (**the signal peptide**) at the amino terminus of the nascent polypeptide. All proteins destined for secretion, inclusion in lysosomes, or insertion into various membranes are made by bound ribosomes reading mRNAs with similar, but non-identical signal sequences. Ribosomes that associate with such mRNAs begin protein synthesis while free in the cytoplasm. As synthesis proceeds, the amino-terminal signal peptide emerges from the large ribosomal subunit and becomes associated with the **signal recognition particle** (SRP), a complex containing several proteins and at least one RNA (7S RNA). Once the SRP is bound to the nascent polypeptide, elongation of the growing protein is blocked unless/until the SRP binds to a protein, called the **docking protein,** in the membrane of the rER. Once the ribosomes (part of a polysome) bind to the membrane, the SRP is released and the large ribosomal subunit (contacting the membrane) induces the association of two additional membrane proteins, **ribophorins I** and **II,** to form a "channel" in the membrane through which the nascent polypeptide moves as protein synthesis proceeds. The signal sequence of hydrophobic amino acids is cleaved from the growing polypeptide (often while synthesis is continuing) by the **signal peptidase** in the lumen of the rough ER. By this mechanism, almost all proteins that must be inserted into or passed through a membrane (such as collagen, serum albumin, the ribophorins, signal peptidase, mannose 6-phosphate receptors, etc.) are synthesized on the rER, while proteins made for usage within the cytoplasm (including actin, myosin, and components of the ribosomes) are made on free ribosomes and thus kept separate from proteins made on the rER.

Proteins made on the rough ER are subject to a number of modifications, some of these being **co-translational** (i.e., while the polypeptide is still bound to the ribosome), while most are **post-translational**. Co-translational modifications include, in addition to removal of the signal peptide by signal peptidase, catalytic rearrangement of disulfide bonds, hydroxylation of proline residues (in collagen), and some of the initial steps in glycosylation (addition of carbohydrate groups to specific amino acid side chains).

Glycosylation of proteins is a complex process that begins in the lumen of the rER, with the addition of a core chain of about 14 sugar residues to the NH_2 side chain of an asparagine residue in the protein. The mannose rich core chain is initially coupled (via a high energy pyrophosphate bond) to a specific membrane lipid - **dolichol**. Transfer of this core from the dolichol to the appropriate asparagine side chain on the protein occurs in the lumen of the rER. Many proteins destined for secretion, accumulation in lysosomes, or insertion into membranes undergo additional glycosylation as a post-translational modification. Much of this glycosylation, while dependent on addition of the core oligosaccharide in the rER, is carried out by specific enzymes in various cisternae of the Golgi apparatus.

Golgi associated enzymes exist that add specific sugars, as well as ones that remove specific sugars. Distinct from **N-linked sugar addition** begun in the rER, there are a number of carbohydrate side chains that are added to hydroxyl groups of serine, threonine or hydroxylysine residues. Such **O-linked oligosaccharide addition** is carried out in the Golgi apparatus. One of the most important functions of the Golgi apparatus is the sorting of polypeptides made in the rER, but destined for use in different places. Hydrolytic enzymes destined for storage in lysosomes have mannose 6-phosphate groups added to N-linked oligosaccharide chains. This addition is carried out by enzymes in cis-Golgi cisternae. A complementary receptor for mannose 6-phosphate groups is found in Golgi membranes and plays a role in the sorting of enzymes destined for packaging in lysosomes from proteins that will not be segregated into lysosomes.

Not all proteins destined for incorporation into membrane bounded organelles are synthesized by ribosomes of the rER. Most of the proteins that make up peroxisomes are synthesized by free cytoplasmic ribosomes. Similarly, many mitochondrial proteins are made on free cytoplasmic ribosomes. Many of the mitochondrial and peroxisomal proteins that are synthesized by free ribosomes are carried to their destinations by **molecular chaperones**, proteins that function to prevent the final folding of such proteins before they reach the target organelle.

Examine the transmission electron micrograph below (**Figure 4.2**) and then choose the one correct lettered answer for each item.

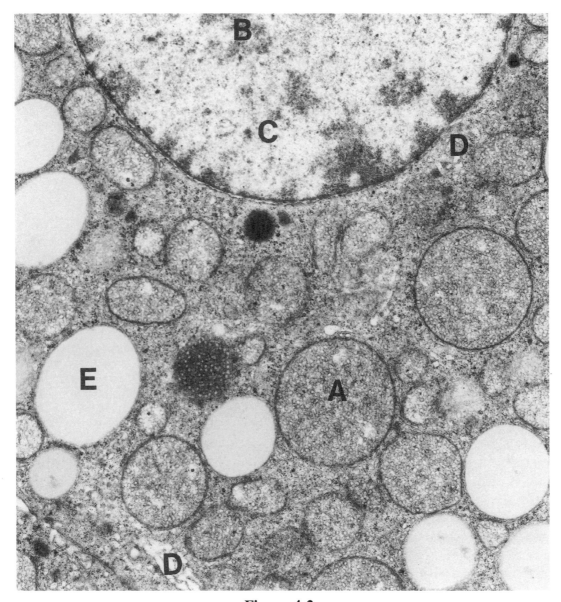

Figure 4.2

330. The letter C is

(A) next to a cluster of ribosomes
(B) on the nuclear envelope
(C) in the lumen of the rough endoplasmic reticulum
(D) on euchromatin
(E) in the matrix of a mitochondrion

331. The structure labelled E is a

    (A)   lipid droplet
    (B)   Golgi apparatus
    (C)   mitochondrion
    (D)   peroxisome
    (E)   lysosome

332. Which physiological function is most characteristic of a cell with this ultrastructure?

    (A)   production of red blood cells
    (B)   steroid synthesis
    (C)   storage of polypeptide hormones
    (D)   glycogen storage
    (E)   motility

333. Which labelled structure is the site of storage of cholesterol esters, precursors for the principal synthetic pathway carried out by this cell?

334. Which labelled organelle produces ATP **and** is involved in the principal synthetic pathway of this cell?

ANSWERS AND TUTORIAL ON ITEMS 330-334

The answers are: **330-D; 331-A; 332-B; 333-E; 334-A. Figure 4.2** is a transmission electron micrograph of a cell from the **adrenal cortex**. It has an ultrastructure characteristic of cells secreting steroids including an abundance of **smooth endoplasmic reticulum** (D) (not well illustrated here), **numerous lipid droplets** (E), large round **mitochondria with tubulovesicular cristae** (A) and small dense bodies. The **nucleolus** (B) in a **nucleus** with much **euchromatin** (C) is prominent. This cell synthesizes cortisol from cholesterol esters stored in lipid droplets (E). **Cholesterol** is released from the lipid droplets and enters mitochondria where it is converted into pregnenolone. In the smooth endoplasmic reticulum, pregnenolone is converted to progesterone and then 17-deoxycorticosterone, which is finally converted into cortisol by mitochondrial enzymes. The factors controlling the shuttling of different intermediates from lipid droplets to mitochondria to smooth endoplasmic reticulum and back to mitochondria are not well understood. Also, the mechanism of steroid secretion is controversial with most authors favoring direct release of steroids by diffusion. Steroid secreting cells are abundant in the adrenal cortex, in the corpus luteum of the ovary, in the placenta and in the interstitium of the testis (Leydig cells).

Examine the following electron micrograph (**Figure 4.3**) which is a transmission electron micrograph of the apical portion of an epithelial cell lining the lumen of the small intestine. Each item either describes the function of a structure labelled in **Figure 4.3** or relates to some aspect of the process in which this cell takes part. For each numbered item choose the labelled structure (A-E) **OR** lettered answer (F-M) that best corresponds.

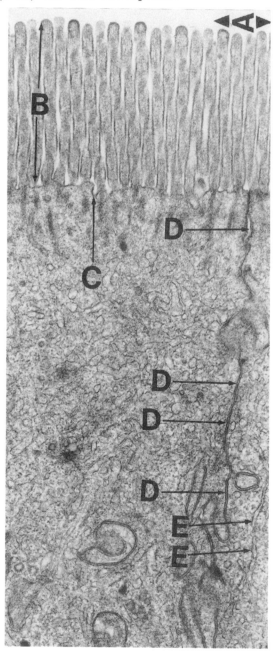

Figure 4.3

(F)   Smooth endoplasmic reticulum
(G)   Chylomicron
(H)   Golgi apparatus
(I)   Lacteal
(J)   Facilitated diffusion
(K)   Simple diffusion
(L)   Receptor-mediated endocytosis
(M)   Ribosomes

335.   Breakdown of the contents of a fat-rich meal occurs here, aided by lipases.

336.   Uptake of monoglycerides and free fatty acids occurs across the apical plasma membrane covering which structures?

337.   The uptake of monoglycerides and free fatty acids occurs by which process?

338.   Which structures contain a core bundle of actin microfilaments?

339.   Where are monoglycerides and free fatty acids recombined to form triglycerides?

340.   Assembly of chylomicrons from proteins, triglycerides and glycolipids occurs here.

341.   Puromycin is most likely to bind to which structures?

342.   The protein components of chylomicrons are produced here.

343.   Chylomicrons are secreted into which space?

344.   Once secreted, chylomicrons are collected in which structure?

ANSWERS AND TUTORIAL ON ITEMS 335-344

The answers are: **335-A; 336-B; 337-K; 338-B; 339-F; 340-H; 341-M; 342-E; 343-D; 344-I**. **Figure 4.3** shows an intestinal epithelial cell. Dietary fats, composed mainly of triglycerides, are hydrolyzed to fatty acids and monoglycerides in the **lumen of the small intestine** (A) by the action of **lipases** secreted by the pancreas. Fatty acids and monoglycerides diffuse across the plasma membranes of intestinal **microvilli** (B) by simple diffusion (L). The microvilli, small projections supported by a core bundle of actin microfilaments, provide an extensive area across which such diffusion can occur. The fatty acids and monoglycerides diffuse into the lumen of elements of the **smooth endoplasmic reticulum** (G) were they are resynthesized into triglycerides. Next, the triglycerides are transported to the **Golgi apparatus** (I) where they are further processed by addition of glycolipids and proteins to form **chylomicrons**. The proteins of chylomicrons are made by the **rough endoplasmic reticulum** (E); this ribosome-dependent protein synthesis can be inhibited by **puromycin**, in which case formation (and thus secretion) of chylomicrons is blocked. Chylomicrons are released from the epithelial cells by secretion across the basolateral plasma membrane from the lateral borders of absorptive epithelial cells into **intercellular spaces** (D). From there they move across the epithelial basement membrane and finally enter the lumen of blind ending lymphatic capillaries called **lacteals** (J). Lacteals anastomose to form larger lymphatic vessels which conduct the chylomicrons to the systemic circulation.

Examine **Figure 4.4** (below) which is a high power transmission electron micrograph of the junctions between three liver parenchymal cells; for each numbered item choose the lettered structure that is most appropriate.

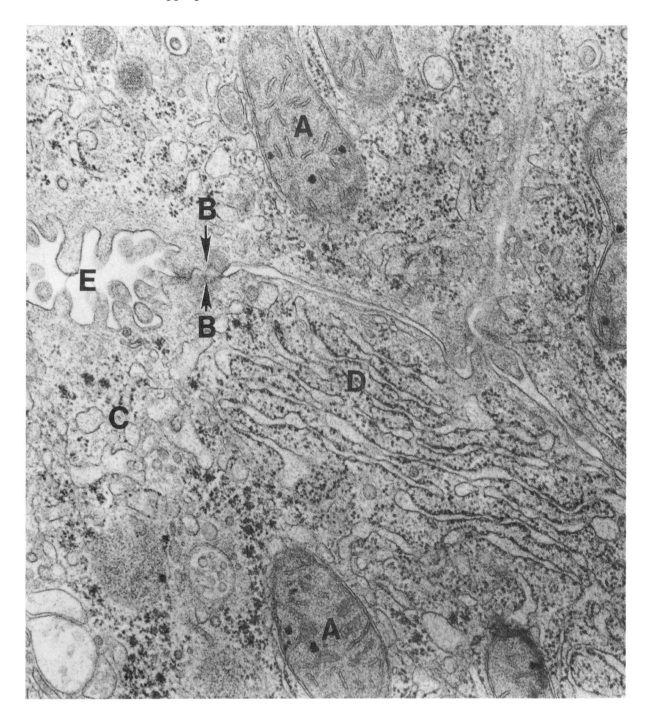

Figure 4.4

345. This structure contains enzymes involved in the electron transport chain and oxidative phosphorylation.

346. This structure forms a tight junction between two hepatocytes, thus functioning as the blood-bile barrier.

347. This is involved in the initial stages of the synthesis of serum albumin.

348. Bile produced by hepatocytes is released into this structure and carried away from the hepatocytes.

ANSWERS AND TUTORIAL ON ITEMS 345-348

The answers are: **345-A; 346-B; 347-D; 348-E**. **Figure 4.4** is a high power transmission electron micrograph showing parts of three adjacent liver parenchymal cells (**hepatocytes**). **Mitochondria** (A) have enzymes of the electron transport chain and oxidative phosphorylation. These enzymes are used to synthesize ATP. This energy rich compound is utilized for many of the anabolic processes occurring in the liver. For example, the liver is actively engaged in protein synthesis. The **rough endoplasmic reticulum** (D) is the site where mRNAs are translated into polypeptide chains. Bile constituents are conjugated in the **smooth endoplasmic reticulum** (C) and are secreted into the **bile canaliculi** (E) between parenchymal cells. **Tight junctions** (B) surround the bile canaliculi and prevent bile from leaking into the vascular spaces in the liver. Thus, these tight junctions represent the anatomical basis for the blood-bile barrier.

Items 349-356 pertain to certain labeled structures in **Figure 4.5** below. Examine the transmission electron micrograph and then choose the best response to answer the items below.

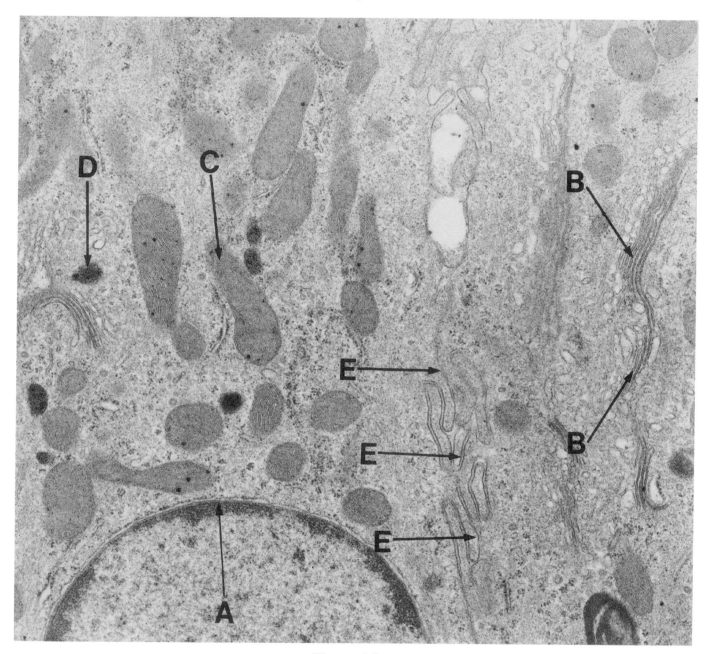

Figure 4.5

349. The structure labeled A is rich in which important protein?

  (A) actin
  (B) clathrin
  (C) dynein
  (D) lamin
  (E) tubulin

350. The structure labeled A is best described as

  (A) peripheral heterochromatin
  (B) intermediate filaments of a desmosome
  (C) an adhesion plaque
  (D) a zonula occludens
  (E) glycogen deposits

351. Which enzyme is characteristic of the structure labeled B?

  (A) trypsin
  (B) DNAse
  (C) DNA polymerase
  (D) galactosyltransferase
  (E) glucuronidase

352. The structure labeled B is **LEAST ABUNDANT** in which cell type?

  (A) goblet cell
  (B) pancreatic acinar cell
  (C) CNS neuron cell body
  (D) PNS neuron cell body
  (E) skeletal muscle

353. Which cell type below would have the **GREATEST ABUNDANCE** of structure C?

  (A) skeletal muscle
  (B) smooth muscle
  (C) cardiac muscle
  (D) erythrocyte
  (E) oligodendroglial cell

354. The structure labeled D would have which function?

(A) storage form of glucose
(B) generation of ATP
(C) synthesis of ribosomes
(D) breakdown of phagocytosed material
(E) synthesis of glycoproteins

355. The structure labeled D is a

(A) glycogen granule
(B) mitochondrion
(C) nucleolus
(D) lysosome
(E) polysome

356. The structure labeled E has which function?

(A) intercellular adhesion
(B) cell movement
(C) storage of Ca^{2+}
(D) storage of genome
(E) movement of chromosomes

ANSWERS AND TUTORIAL ON ITEMS 349-356

The answers are: **349-D; 350-A; 351-D; 352-E; 353-C; 354-D; 355-D; 356-A**. In **Figure 4.5**, the structure labelled A is the material adhering to the internal (nucleoplasmic) surface of the nuclear envelope; it is sometimes referred to as the cytoskeleton of the nuclear envelope because it contains a variety of proteins that function to stabilize the envelope and to maintain its interaction with the heterochromatin (condensed chromatin) immediately subjacent. In addition to the DNA and proteins of the peripheral heterochromatin, the domain labelled A is rich in nuclear **lamins** - proteins with sequence homology to the proteins that form cytoplasmic intermediate filaments. The structure labelled B is the **Golgi apparatus**, a stack of flattened membranes and vesicles that contain many enzymes - such as **galactosyltransferase** - that participate in the addition of carbohydrate moieties to recently synthesized proteins. The Golgi apparatus is highly developed in all cells (such as goblet cells and pancreatic acinar cells) that secrete proteins, especially those that are heavily glycosylated. Because the Golgi is also involved in routing newly-synthesized proteins to specific parts of cells it is also prominent in neurons. Although skeletal muscle cells contain a Golgi apparatus, it is quite small and relatively unimportant in muscle cells. The structure labelled C is a **mitochondrion**. As the principal site of ATP production via oxidative phosphorylation, mitochondria are found in virtually all eukaryotic cells. They are in greatest abundance in cells that utilize large amounts of ATP; of the cells listed in Item 353, cardiac muscle cells require large amounts of ATP and would have **many** mitochondria. The dense, membrane-delimited structure labelled D is a **lysosome**. These organelles are packages containing a variety of hydrolytic enzymes. The proteases, nucleases and other such degradative enzymes they contain are used to break down extracellular materials that are phagocytosed by cells. Label E indicates where the plasma membrane of the cell is tightly joined to the neighboring cell by an **adherent junction**.

Examine the scanning electron micrograph (**Figure 4.6**) below, which shows two of the cells normally found in peripheral blood, and then answer the items with the **ONE** best response.

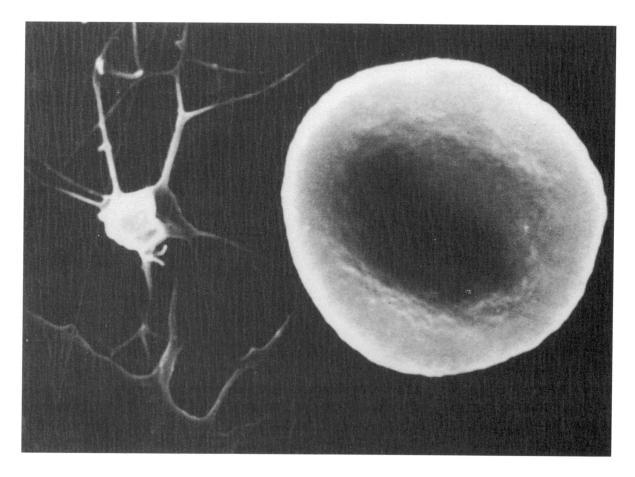

Figure 4.6

357. The rounder structure on the right is a

    (A)   red blood cell
    (B)   eosinophil
    (C)   basophil
    (D)   platelet
    (E)   lymphocyte

358.   The stellate structure on the left is a

    (A)   red blood cell
    (B)   eosinophil
    (C)   basophil
    (D)   platelet
    (E)   lymphocyte

Items 359-366 should be answered A-D from the list below.

    (A)   Rounder structure on the right
    (B)   Stellate structure on the left
    (C)   Both
    (D)   Neither

359.   contains a nucleus surrounded by a nuclear envelope

360.   has endoplasmic reticulum and lysosomes

361.   has cytoplasmic microtubules

362.   is rich in spectrin

363.   derived from bone marrow

364.   part of the mononuclear phagocyte system

365.   is capable of mitosis

366.   is morphologically abnormal in hereditary spherocytosis

ANSWERS AND TUTORIAL ON ITEMS 357-366

The answers are: **357-A; 358-D; 359-D; 360-B; 361-B; 362-A; 363-C; 364-D; 365-D; 366-A**. **Figure 4.6** is a scanning electron micrograph of human **peripheral blood**. The cell on the right is an **erythrocyte**, the only cell that, under normal conditions, has the shape of a biconcave disk. The structure on the left is a **platelet**, the only one of the choices in Item 358 that is smaller than a red blood cell. Neither of these has a nucleus; hence neither is capable of mitosis. The erythrocyte nucleus is extruded during maturation along with essentially all other organelles. The erythrocyte is rich in **spectrin**, a peripheral membrane protein that is defective in **hereditary spherocytosis** and leads to structural abnormalities in erythrocytes.

The platelet is a fragment of a **megakaryocyte** and does not contain a nucleus. While the erythrocyte is nearly devoid of intracytoplasmic organelles, the platelet has a few mitochondria, some granules and some vacuoles as well as a well defined peripheral bundle of microtubules.

Both erythrocytes and platelets are derived from bone marrow stem cells. The former is derived from the erythroid series of stem cells (CFU-E forms erythroblast which forms erythrocytes) and the latter is derived from CFU-M which forms a megakaryocyte, which fragments to form many platelets. Neither cell type is part of the mononuclear phagocyte system.

Examine the following diagrammatic view (**Figure 4.7**) of a cell as seen in the transmission electron microscope; the black material in the upper left of the diagram is calcified extracellular matrix.

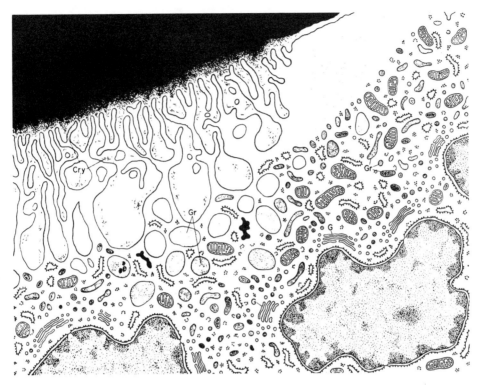

Figure 4.7

367.   This cell is a/an

   (A)   bone marrow stem cell
   (B)   osteoblast
   (C)   fibroblast
   (D)   osteoclast
   (E)   nerve cell

368.   The highly irregular surface of this cell is best described as

   (A)   microvilli
   (B)   stereocilia
   (C)   a ruffled membrane
   (D)   canaliculi
   (E)   fascia adherens

369. This cell is **LEAST LIKELY** to be involved in

   (A)   active transport of Cl⁻ out of the cell
   (B)   creation of an acidic environment to solubilize Ca^{2+} from mineralized matrix
   (C)   phagocytosis
   (D)   proliferation
   (E)   motility

370. Which of the following other cell types in the human body is most similar in structure and function to the cell shown in **Figure 4.7**?

   (A)   plasma cell
   (B)   erythrocyte
   (C)   goblet cell
   (D)   CNS neuron
   (E)   gastric parietal cell

ANSWERS AND TUTORIAL ON ITEMS 367-370

The answers are: **367-D; 368-C; 369-D; 370-E. Figure 4.7** shows a large multinucleate cell called an **osteoclast**. Osteoclasts reside on the surfaces of bone spicules and, under the stimulus of parathyroid hormone, they break down the bone. This degradative activity involves the formation of an extensively folded surface (ruffled membrane) in contact with the bone surface. The osteoclasts secrete hydrogen ions across this folded surface, coupled to the release of chloride ions. The acidification of the immediate environment of the bone matrix leads to the decalcification of the matrix, thus increasing the calcium level in the blood stream. Osteoclasts also break down the proteinaceous bone matrix, by phagocytosing (also across the ruffled membrane border) chunks of the matrix and degrading them using the hydrolytic enzyme contents of lysosomes. Osteoclasts are end-stage cells and do not proliferate. There are other cells in the body that have functions (and thus structures) that are similar to those of the osteoclasts. Of the list of cells in Item 370, the one with the greatest such similarities to osteoclasts is the **gastric parietal cell**; this cell also has an extensively folded surface (the apical surface in the case of the parietal cell, which is an epithelial cell) across which hydrogen ions are actively secreted, thus acidifying the adjacent extracellular space (the lumen of the stomach).

Examine the high power transmission electron micrograph in **Figure 4.8** below and then provide the best response to the items that follow.

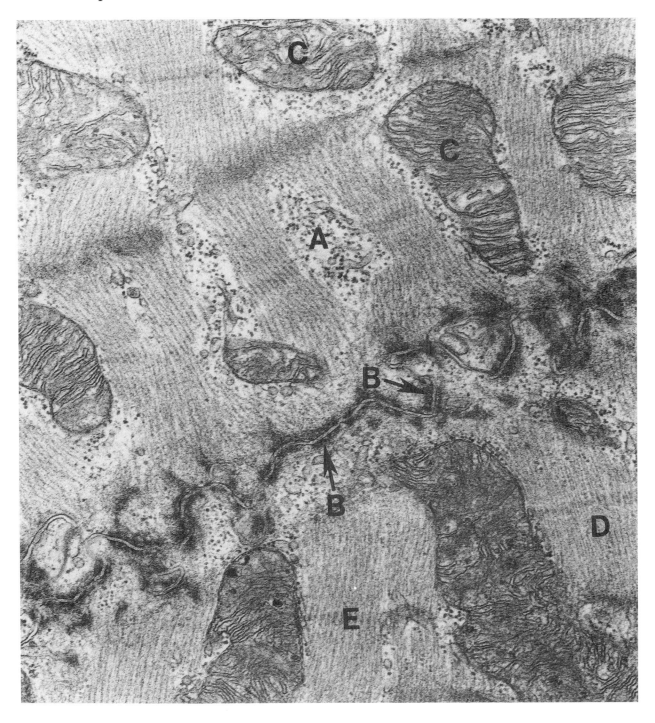

Figure 4.8

371. The organelles labeled C are

   (A)   nuclei
   (B)   mitochondria
   (C)   lysosomes
   (D)   pinocytotic vesicles
   (E)   ribosomes

372. The chief function of these organelles is

   (A)   contains nuclear genome
   (B)   oxidative phosphorylation
   (C)   phagocytosis
   (D)   receptor-mediated endocytosis
   (E)   protein synthesis

373. Which of the following enzymes is most prominent in these organelles?

   (A)   RNA polymerase
   (B)   DNA ligase
   (C)   ATP synthase
   (D)   glucuronidase
   (E)   hyaluronidase

374. The dark spots around the letter A are

   (A)   ribosomes
   (B)   pigment granules
   (C)   glycogen deposits
   (D)   calcium storage masses
   (E)   ferritin particles

375. These cells are most likely

   (A)   fibroblasts
   (B)   cardiac muscle cells
   (C)   plasma cells
   (D)   erythrocytes
   (E)   smooth muscle cells

376. The masses indicated at the letters B are

  (A)   ribosomes of the rough endoplasmic reticulum
  (B)   desmosomes
  (C)   keratin granules
  (D)   postsynaptic densities of neuromuscular junctions
  (E)   microtubule organizing centers

ANSWERS AND TUTORIAL ON ITEMS 371-376

The answers are: **371-B; 372-B; 373-C; 374-C; 375-B; 376-B. Figure 4.8** is a transmission electron micrograph of portions of two **cardiac muscle cells**. Much of the field is filled with the **thick (myosin-rich) and thin (actin-rich) filaments** of the cardiac myofibrils. The organelles labeled C are **mitochondria**. Mitochondria carry out **oxidative phosphorylation**; one of the essential enzymes they contain is **ATP synthase**, which uses the potential energy derived from the breakdown of foodstuffs to produce the high energy compound ATP. The constant contractile activity of the cardiac cells, driven by the cyclic hydrolysis of ATP by the myosin and actin of the myofibrils, requires a rich supply of ATP from the mitochondria and the rapid replenishment of that supply via the break down of **glycogen granules** (A) - a site at which glucose is stored in the form of glycogen polymers. The contraction of adjacent cardiac muscle cells is coordinated across extensive **gap junctional contacts** which are held in place because the cells are adherent at numerous places (fascia adherens) containing **desmosomes**, the dark, membrane associated structures labelled B in the micrograph.

Examine the high magnification electron micrograph shown in **Figure 4.9** and, for each item, select the most appropriate answer.

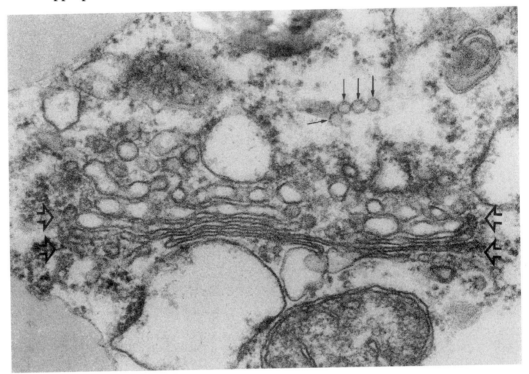

Figure 4.9

377.   The large organelle between the large open arrows is

(A)   the Golgi apparatus
(B)   a mitochondrion
(C)   part of the rough endoplasmic reticulum
(D)   a portion of the nuclear envelope
(E)   a secondary lysosome

378.   This organelle is **MOST** likely to be involved in

(A)   production of steroid hormones
(B)   transcription of DNA into RNA
(C)   synthesis of ribosomal RNA
(D)   addition of carbohydrate to secretory proteins
(E)   routing of actin to the place where microfilaments are assembled

379. The small vesicles at the 4 small arrows

    (A)   carry transfer RNA to the rough endoplasmic reticulum
    (B)   shuttle materials from the rough endoplasmic reticulum to the Golgi apparatus
    (C)   contain proteins destined to be incorporated into mitochondria
    (D)   are the portions of the smooth endoplasmic reticulum that store calcium
    (E)   are tertiary lysosomes

380. Which completion of the sentence makes the **LEAST** accurate statement about the Golgi apparatus? The Golgi apparatus

    (A)   is found in most cells but is **least** pronounced in cells that secrete glycoproteins.
    (B)   contains membrane-delimited cisternae and vesicles.
    (C)   serves as a routing center that directs proteins made on the rough endoplasmic reticulum to their most likely place of utilization.
    (D)   is often found near the nucleus.
    (E)   contains enzymes that are non-uniformly distributed amongst the various membrane cisternae.

ANSWERS AND TUTORIAL ON ITEMS 377-380

The answers are: **377-A; 378-D; 379-B; 380-A. Figure 4.9** is a high magnification of a cell containing a prominent **Golgi apparatus** (between the large arrows). The Golgi apparatus consists of a stack of flattened membranous sacs (cisternae) surrounded by a "cloud" of small membrane vesicles; the assembly is usually close to the cell nucleus. The Golgi apparatus functions in the **post-translational modification of proteins** synthesized by the ribosomes of the rough endoplasmic reticulum. One of the most important of these modifications is the addition of carbohydrates to glycoproteins; hence the Golgi is very highly developed in cells - such as goblet cells - that secrete heavily glycosylated proteins. The Golgi is also involved in the routing of proteins made in the rER to their various destinations, including **lysosomes, zymogen storage granules,** and different portions of the **plasma membrane.** The small vesicles (small arrows) at the periphery of the Golgi serve to transfer proteins from the rER to the **cis-Golgi saccules,** from the cis- to the **trans-saccules,** and from the trans-side of the Golgi to specific destinations such as the apical plasma membrane, basal-lateral plasma membrane, etc. The various Golgi saccules contain subsets of the enzymes in the Golgi; thus proteins being processed through the Golgi are acted upon sequentially by the various processing enzymes.

381. Which of the following processes does **not** involve transferring material(s) from one side of the plasma membrane to the other?

    (A) exocytosis
    (B) phagocytosis
    (C) receptor-mediated endocytosis
    (D) pinocytosis
    (E) autophagocytosis

382. Exocytosis

    (A) is the name given to the removal of asparagine from a polypeptide chain.
    (B) involves fusion of the membrane of a vesicle with the plasma membrane.
    (C) only occurs across the apical surface of an epithelial cell.
    (D) can only occur if the membranes around secretory granules have the same polypeptide composition as the plasma membrane.
    (E) leads to the release of vesicle membranes into the extracellular space.

383. All of the following are features of receptor-mediated endocytosis **EXCEPT**:

    (A) Uptake of extracellular material by formation of an invagination and pinching off of the plasma membrane.
    (B) Binding of material, destined for import into the cell, to a receptor accessible on the outside surface of the plasma membrane.
    (C) Non-selective uptake of lipid soluble molecules.
    (D) Removal of some receptor proteins from the plasma membrane.
    (E) Internalization of some infectious agents such as bacteria and viruses.

384. All of the following events are usually a part of phagocytosis **EXCEPT**:

    (A) Uptake of some extracellular fluid in a non-specific fashion.
    (B) Release of lysosomal enzymes into the extracellular space.
    (C) Binding of extracellular materials to the plasma membrane in a process that does **not** involve hydrolysis of ATP.
    (D) Internalization of materials that will subsequently be found in the lumen of a secondary lysosome.
    (E) Removal from the plasma membrane of portions of the membrane.

385. Regulated exocytosis

  (A)   occurs when the secreting cell is depleted of membranes required for constitutive exocytosis.
  (B)   can only occur at portions of the plasma membrane adjacent to a basal body.
  (C)   is a receptor-mediated event by which a stimulus leads to selective release of materials stored in membrane vesicles.
  (D)   only occurs from cells incapable of constitutive secretion.
  (E)   is the process by which membrane proteins taken up during endocytosis are returned to the plasma membrane.

ANSWERS AND TUTORIAL TO ITEMS 381-385

The answers are: **381-E; 382-B; 383-C; 384-B; 385-C**. A number of cellular processes involve the controlled uptake (**endocytosis**) or release (**exocytosis**) of materials within membrane bounded spaces. Exocytosis is most frequently the mode by which materials, synthesized within a cell and transiently stored in membrane vesicles, are released. During exocytosis the membrane of the vesicle fuses with the plasma membrane of the cell and the vesicle contents are released into extracellular space. The vesicle membrane becomes - at least transiently - part of the plasma membrane but may be recaptured as part of a subsequent endocytotic event. Exocytosis may be in response to a specific signal (regulated exocytosis) or occur at some basal rate without any specific stimulus (constitutive exocytosis). Cells such as the goblet cells lining portions of the gastrointestinal tract carry out both constitutive and regulated exocytosis. When cells take up materials from the extracellular milieu by pinching off portions of the plasma membrane, they are engaged in endocytosis. If the vesicles carry fluid with little or no selectivity, they are referred to as pinocytotic vesicles. Large vesicles - containing bacteria, cells (or portions thereof), or extracellular debris - are involved in phagocytosis. There is a grey area in which the materials that are taken up are macromolecular complexes (e.g., viruses), but not large enough to be visualized by conventional light microscopy; such endocytotic events may or may not be referred to as phagocytosis. In virtually all cases of phagocytosis, the material to be transported is bound to some component of the plasma membrane in a process that does not require energy input (proceeds at low temperature); subsequent events - requiring energy - involve the pinching off of a portion of the plasma membrane to form a vesicle containing the bound material and, usually, some of the extracellular fluid that is internalized along with the macromolecular material. Materials taken up by phagocytosis are usually - but not always - degraded in a process that involves fusion of the **phagocytic vesicle** (phagosome) with the membrane of a **lysosome**, thus forming a **phagolysosome** (secondary lysosome). In receptor-mediated endocytosis, certain molecules or macromolecular complexes are tightly bound to specific receptors and induce the formation of endocytic vesicles (**endosomes**) that may/may not convey the ingested materials to the lysosomal degradation pathway.

This set of items is related to the process of ribosome synthesis and assembly. For each numbered item choose the lettered answer that is most appropriate.

386. The nucleolus

(A) is tightly bounded by and adherent to the inner membrane of the nuclear envelope.
(B) contains the chromosomal regions needed for synthesis of all ribosomal RNAs.
(C) functions as the domain where ribosomal proteins and nucleic acids are made.
(D) acts to synthesize most ribosomal RNAs and to assemble ribosomes.
(E) is usually disassembled during S phase of the cell cycle.

387. The granular portions of the nucleolus are most likely involved in

(A) synthesis of 5S ribosomal RNA.
(B) assembly of ribosomal proteins and nucleic acids.
(C) formation of the complex of mRNA and the small ribosomal subunit that is required for initiation of protein synthesis.
(D) attachment of nucleolar organizers to the nuclear envelope.
(E) packaging of ribosomal RNAs into nucleosomes.

388. RNA polymerase I

(A) is most concentrated in heterochromatic regions of the nucleus.
(B) synthesizes 45S pre-ribosomal RNA, which is then processed to yield 28S, 18S and 5.8S ribosomal RNAs.
(C) transcribes 5S ribosomal RNA from 5S RNA genes found throughout euchromatin regions.
(D) is responsible for the production of the mRNAs needed to code for ribosomal proteins.
(E) functions in the formation of large ribosomal subunits, but not in the synthesis of the small subunits.

389. 5S ribosomal RNA

(A) is encoded by genes that form the fibrillar region of the nucleolus.
(B) forms part of the small ribosomal subunit.
(C) is transcribed by RNA polymerase III.
(D) must be processed by topoisomerase before it can be incorporated into mature ribosomes.
(E) is cleaved from the 45S ribosomal RNA precursor by enzymes found associated with the nuclear envelope.

390. Passage of materials through nuclear pores is **LEAST** likely to be involved in which of the following?

    (A)   Transport of the mRNAs encoding ribosomal proteins from the nucleus to the cytoplasm.

    (B)   Passage of ribosomal proteins from the cytoplasm into the nucleolus.

    (C)   Transit of mature ribosomal subunits from the nucleolus to the cytoplasm.

    (D)   Transit of precursor nucleotides from the cytoplasm into the nucleus.

    (E)   Passage of 5S ribosomal RNA from euchromatin into the nucleolus.

ANSWERS AND TUTORIAL ON ITEMS 386-390

The answers are: **386-D; 387-B; 388-B; 389-C; 390-E**. The **nucleolus** is a specialized domain within the nucleus of most actively growing or protein synthesizing eucaryotic cells (**Figure 4.10**).

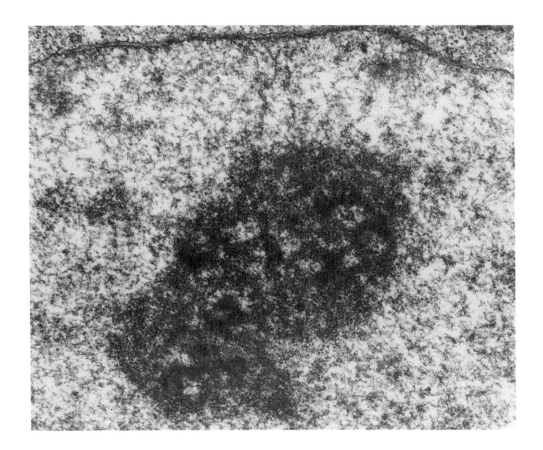

Figure 4.10

It is usually found within the general chromatin, with no obvious attachment to the nuclear envelope. Specialized regions (**nucleolar organizer regions**) of several chromosomes come together to form the nucleolus, which disassembles at the start of mitosis and is reformed after telophase. The nucleolus has a **fibrillar region**, where ribosomal RNAs are synthesized and a **granular region** where ribosomal subunits are assembled. Eucaryotic cells contain three distinct enzymes that transcribe DNA sequences into RNA. Most messenger RNAs (mRNAs) coding for proteins are transcribed by **RNA polymerase II**. The mRNAs coding for the approximately 80 ribosomal proteins are produced by **RNA polymerase II**. Ribosomal proteins are synthesized in the cytoplasm by free ribosomes and transported into the nucleus via passage through **nuclear pores**. They are assembled into ribosomal subunits, in combination with ribosomal RNAs, in the granular portions of the nucleolus. Assembled ribosomal subunits exit the nucleus via nuclear pores, thus reaching the cytoplasm, where they participate in protein synthesis. The RNAs that are required for the assembly of ribosomal subunits are made by two distinct pathways. The 5S RNA that is found in the large ribosomal subunit is encoded by genes found **outside** of the nucleolus. These genes are transcribed by RNA polymerase III and are transported to the granular regions of the nucleolus where subunit assembly occurs. The chromosomal genes that encode the 28S, 18S and 5.8S ribosomal RNA are found in the nucleolar organizer regions and are transcribed by RNA polymerase I. Their transcription produces a 45S pre-ribosomal RNA that, via a series of RNA processing steps, is cleaved to produce the final 18S RNA of the small ribosomal subunit and the 28S and 5.8S RNAs of the large subunit.

CHAPTER V
REGULATION OF CELLULAR FUNCTION

Items 391-394

Insulin is a polypeptide hormone secreted by specialized cells in the pancreatic islets of Langerhans. The following set of items pertains to the structure and physiological actions of insulin and its receptor.

391.   All of the following statements concerning insulin structure are true **EXCEPT**:

    (A)   insulin monomer has a MW = 12,000
    (B)   preproinsulin in synthesized on the endoplasmic reticulum of β cells
    (C)   proinsulin is stored in the Golgi apparatus after cleavage of a peptide from preproinsulin
    (D)   zinc is associated with circulating insulin polymers
    (E)   increased intracellular glucose stimulates insulin release from β cells

392.   All of the following statements concerning insulin receptors are true **EXCEPT**:

    (A)   they are found on liver and fat cell membranes
    (B)   they are clustered on fat cell membranes
    (C)   they have an extracellular insulin binding subunit
    (D)   they are clustered on the membranes of liver and fat cells
    (E)   they are internalized once insulin is bound

393.   All of the following are true statements concerning the effects of insulin **EXCEPT**:

    (A)   it increases glycogen synthesis in liver cells
    (B)   it binds to the α subunit of the insulin receptor
    (C)   it cannot exert its effects unless internalized into cells
    (D)   it stimulates tyrosine kinase activity of the β subunit
    (E)   it decreases intracellular glucose transport

394. All of the following are true statements concerning insulin effects on glycogen breakdown and gluconeogenesis **EXCEPT**:

(A) it increases glycogen synthesis in muscle cells
(B) it increases synthesis of fatty acids in fat cells
(C) it decreases synthesis of triacylglycerol in liver cells
(D) it decreases gluconeogenesis in liver cells
(E) it promotes glycolysis and ATP synthesis

ANSWERS AND TUTORIAL ON ITEMS 391-394

The answers are: **391-A; 392-D; 393-C; 394-C**. The **insulin monomer** (MW = 6,000) consists of an **A chain** of 21 amino acids and a **B chain** of 30 amino acids. These two chains are joined together by -S-S- bridges. β cells in the pancreatic islets of Langerhans synthesize **preproinsulin** (with a leader sequence of 16 amino acids) on the rough endoplasmic reticulum. The leader sequence is cleaved by proteolysis to form the proinsulin molecule which contains an A and B chain and a **C peptide** and is stored in the Golgi apparatus. Hydrolysis of arg-lys and arg-arg peptide bonds frees the C peptide from the insulin monomer. Insulin can form polymers of 2 or more monomers at high insulin concentration in association with Zn^{2+}.

The **insulin receptor** is a transmembrane complex that is abundant on the surface of liver and fat cells. It can be isolated and purified by detergent solubilization followed by chromatography on an insulin-affinity column. It is a tetramer of 2 α (MW = 135,000) and 2 β (MW = 90,000) monomers. The α subunits are located on the external side of the cell membrane and have a recognition site for insulin. Insulin binding to receptor even without internalization of hormone-receptor complexes can have profound effects on cells because of conformational changes that are induced by insulin binding. The β subunits have three domains 1) for interaction with α subunits 2) a **transmembrane domain** and 3) a **cytoplasmic domain** with **tyrosine kinase activity**. Autophosphorylation of the β subunit is stimulated directly by insulin binding to the α subunit and is probably involved in the regulation of insulin action.

When glucose concentration rises in blood, increased concentration of glucose in pancreatic β cells causes insulin secretion. When insulin binds to the insulin receptor, it stimulates glucose and amino acid transport and stimulates synthesis of proteins, glycogen, fatty acids and triacylglycerol in liver, muscle and fat cells. Insulin results in the formation of large amounts of ATP by stimulating glycolysis and subsequent oxidation of intermediates in the TCA cycle. Increases in glycogen synthesis, increases in glucose transport, decreases in gluconeogenesis and increases in glycolysis all contribute to the net effect of insulin which is to decrease blood glucose levels.

395. The contraction of smooth muscle involves all of the following biochemical changes **EXCEPT**:

    (A)   release of Ca^{2+} from the sarcoplasmic reticulum
    (B)   phosphorylation of myosin
    (C)   interaction of actin and myosin
    (D)   hydrolysis of ATP
    (E)   binding of Ca^{2+} to calmodulin

396. Which statement is correct concerning biochemical differences between cardiac muscle and skeletal muscle?

    (A)   Cardiac muscle has a myosin with only one heavy chain.
    (B)   The actin in cardiac muscle binds GTP rather than ATP.
    (C)   Tropomyosin isolated from cardiac muscle is almost twice as large as that from skeletal muscle.
    (D)   Cardiac muscle troponin complex does not have a Ca^{2+} binding component (Tn-C).
    (E)   Mitochondria are more numerous in cardiac muscle, reflecting the greater dependence on aerobic metabolism.

ANSWERS AND TUTORIAL ON ITEMS 395 AND 396

The answers are: **395-A; 396-E**. The **thick and thin filament** arrays of **cardiac muscle** are very similar to those of skeletal muscle and the constituent proteins are nearly identical or very closely related. In particular, actin, myosin, tropomyosin and troponin of cardiac muscle differ little from those of skeletal muscle. Cardiac muscle cells contract regularly 24 hours a day for your entire life. They rely nearly exclusively on aerobic metabolism. Imagine the consequences of lactate accumulation in and paralysis of the cardiac muscle. Because of this reliance on aerobic metabolism, mitochondria are much more abundant in cardiac muscle than in skeletal muscle.

    Cardiac muscle does not depend on nerve impulses to trigger contraction, although nerve stimuli do modulate the intrinsic contractile rhythm. **Smooth muscle** cells have numerous thin filaments but, compared with skeletal and cardiac muscle, fewer thick filaments. Myofibrils are not found. Smooth muscle contraction **does** involve hydrolysis of ATP by myosin and actin-myosin interaction resulting in relative sliding of the respective filaments. Smooth muscle cells are stimulated by autonomic innervation or signal transfer from adjacent smooth muscle cells via gap junctions. Their activity is modulated by hormones such as **epinephrine**. Smooth muscle contraction is triggered by Ca^{2+} but there is no regular arrangement of sarcoplasmic reticulum. When smooth muscle cells are stimulated, Ca^{2+} is released from small vesicles lying just beneath the sarcolemma. Ca^{2+} binds to **calmodulin** and the complex then associates with and activates a **myosin light chain kinase**. The kinase phosphorylates one of the myosin light chains. This phosphorylation is required for myosin to interact with actin. Cytoplasmic free Ca^{2+} levels are subsequently reduced by the activity of pumps in the sarcolemma and/or the subsarcolemmal vesicles. This leads to a reduction of the light chain kinase activity and dephosphorylation of the light chain by a myosin light chain kinase.

Items 397-400

A 20 year-old female who is a senior in a state university reports to student health complaining of weakness in both arms and a tendency to fatigue easily while working at her computer. Physical examination reveals that she has trouble keeping her eyes fully opened and her head up. She also complains that, although she has previously jogged regularly, she now finds it difficult to breath comfortably while exercising. She has no siblings but reports that the youngest sister of her mother has had some muscle problems that responded to drug treatments, although the patient could not remember the name of the drug used.

397.   Which of the following is the most likely diagnosis?

(A)   nemaline myopathy
(B)   poliomyelitis
(C)   Duchenne's muscular dystrophy
(D)   myasthenia gravis
(E)   mitochondrial myopathy

398.   This condition is the result of

(A)   an autoimmune response
(B)   excessive exercise
(C)   dietary deficiency in Ca^{2+}
(D)   congenital failure of development of innervation
(E)   a viral infection

399.   Which of the following proteins is most directly affected in this disease?

(A)   acetylcholinesterase
(B)   acetylcholine receptor
(C)   protein kinase C
(D)   calsequestrin
(E)   troponin-C

154

400. Which of the following drugs is most effective in the treatment of this disorder?

    (A)   succinylcholine
    (B)   penicillin
    (C)   cortisone
    (D)   pyridostigmine bromide
    (E)   phenobarbital

ANSWERS AND TUTORIAL ON ITEMS 397-400

The answers are: **397-D; 398-A; 399-B; 400-D**. This woman is suffering from **myasthenia gravis** (MG), an acquired defect that leads to an **autoimmune response** against the **acetylcholine receptor** in the sarcolemma at the neuromuscular junction. Because of this, acetylcholine released from vesicles at the presynaptic membrane does not bind to the receptors and the muscle action potential is not triggered. Common symptoms include muscle weakness, difficulty in raising the arms, drooping eyelids and problems in holding the head erect. MG occurs three times more often in females, with onset around 20 years. In some cases, it can be treated with compounds such as pyridostigmine bromide or neostigmine methyl sulfate that inhibit **acetylcholinesterase** and thus increase the chance that ACh can bind to receptors that are not totally blocked by the auto-antibody.

Various cytokines serve crucial roles in the maturation and stimulation of different cellular subpopulations. Match the cytokine in the answers with the most appropriate description of its primary function in the items below.

(A)	IL-1
(B)	IL-2
(C)	IL-3
(D)	IL-4
(E)	IL-5
(F)	IL-6
(G)	IL-7
(H)	IL-8
(I)	IL-9
(J)	IL-10
(K)	Gamma interferon (INF-γ)
(L)	Transforming growth factor-β

401. Produced by helper T cells, stimulates B cell differentiation and proliferation, also promotes IgA switching.

402. Produced by helper T cells, enhances both class I and class II MHC expression, also promotes immunoglobulin switching to IgG and IgE, in mice.

403. Produced by a many cell types such as B cells, macrophages and monocytes, enhances natural killer cell activation and stimulates helper T cell activation.

404. Produced by macrophages, chemoattractant for neutrophils.

405. Produced by platelets, lymphocytes and macrophages, attracts other macrophages and monocytes, induces IL-1 production.

ANSWERS AND TUTORIAL ON ITEMS 401-405

The answers are: **401-E; 402-D; 403-A; 404-H; 405-L.** The efficiency of the immune response mechanism is dependent upon direct cell-to-cell contact and indirect chemical signaling by cytokines. A large array of cytokines are produced by leucocytes. They are called **interleukins**

(IL); and, along with **gamma interferon** (IFN-γ), are the major soluble mediators of immunological function. An evaluation of the interleukins demonstrates multiple regulatory functions. Interleukins are designated by IL- followed by a number. All are proteins and many are glycosylated. They have molecular weights ranging from 15,00 to 60,000. Most have mitogenic properties and each is antigenically unique. Functionally different interleukins share similar roles in stimulating certain B cell or T cell populations.

IL-1 (A) is produced by many cells and has both T cell and B cell activating functions. It also induces IL-2 receptor expression. IL-1 stimulates the cytocidal actions of cytotoxic T cells, macrophages and natural killer cells. **IL-2** (B) up-regulates lymphokine production and cytotoxic activity of different leucocytes. **IL-3** (C) helps to maintain the growth of mast cells and differentiation of monocytes. **IL-4** (D) enhances both class I and II MHC expression, stimulates IgG4 and IgE production and stimulates IgE receptor expression on B cells. **IL-5** (E) stimulates B cell growth, induces differentiation of eosinophils, stimulates IgA secretion and stimulates B cell growth. **IL-6** (F) stimulates IgG secretion, induces the growth of plasma cells and activates B cells. **IL-7** (G) stimulates B cell differentiation and increases IL-2 receptor expression. **IL-8** (H), produced by macrophages, is a neutrophil chemoattractant. **IL-9** (I) stimulates helper T cells. **IL-10** (J) suppresses production of certain other lymphokines produced by helper T cells. **Transforming growth factor-β** (L) is produced by macrophages, lymphocytes and platelets. It acts to attract other macrophages and monocytes and stimulates IL-1 production. It may also down-regulate the inflammatory response. It is important to realize that cytokines usually have multiple roles which may change under different immunological conditions. Many cytokines also act **synergistically**.

Items 406-415

This set of items deals with basic mechanisms of hormone action. Choose the best answer.

Matching.

    (A)   Membrane permeant (Group I) hormones
    (B)   Membrane impermeant (Group II) hormones
    (C)   Both
    (D)   Neither

406.   Are lipophilic and often derived from cholesterol.

407.   Typically bind to receptors in the nucleus or cytosol.

408. Typically bind to integral membrane proteins.

409. Usually bind to carrier proteins in plasma.

Choose the best response.

410. All of the following hormones activate adenylate cyclase when they bind to receptor **EXCEPT**:

(A) cortisol
(B) ACTH
(C) TSH
(D) CRH
(E) vasopressin

411. The second messenger for vascular smooth muscle relaxation triggered by atrial natriuretic factor is

(A) cAMP
(B) cGMP
(C) calmodulin
(D) inositol triphosphate
(E) arachidonic acid

412. The second messenger for glucagon is

(A) cAMP
(B) cGMP
(C) calmodulin
(D) inositol triphosphate
(E) arachidonic acid

413. The most widespread change in enzymatic activity as a direct effect of increase in cAMP is

(A) increase in protein kinase
(B) decrease in protein kinase
(C) increase in phosphodiesterase
(D) increase in phosphoprotein phosphatase
(E) decrease in phosphoprotein phosphatase

414. All of the following statements concerning G-proteins are true **EXCEPT**:

   (A)   they are integral membrane proteins
   (B)   they mediate adenylate cyclase activity
   (C)   they contain four distinct subunits
   (D)   they can form inhibitory or stimulatory entities
   (E)   they interact with hormone receptors

415. Second messengers have which features in common?

   (A)   similar chemical structure
   (B)   they all activate enzyme cascades
   (C)   they are present in sequestered pools in cells
   (D)   their turnover is very slow
   (E)   lipid solubility

ANSWERS AND TUTORIAL ON ITEMS 406-415

The answers are: **406-A; 407-A; 408-B; 409-A; 410-A; 411-B; 412-A; 413-A; 414-C; 415-B**. **Hormones** bind specifically to **receptors**, located either in the cell membrane or in the cytoplasm. Receptors have high affinity for their respective hormones. Hormones circulate at exceedingly low concentrations in the interstitial fluids, e.g., in the range 10^{-15} - 10^{-9} molar. Receptors have a **recognition domain** that binds the hormone and a **transduction domain** that couples hormone binding and alteration of cellular function.

   Group I hormones are lipophilic, and, except for T_3 and T_4, are derived from cholesterol. These hormones are associated with carrier (transport) proteins in blood. This association circumvents problems with solubility and also prolongs the hormone half-life in plasma ($t_{\frac{1}{2}}$ = days). Free hormone diffuses across the plasma membrane readily and rapidly associates with receptors in the nucleus or cytosol to form a ligand-receptor complex which serves as the intracellular messenger. The hormone-receptor complex now undergoes an activation reaction, rendering it able to bind to the hormone response element of DNA. When the activated ligand-receptor complex binds to the hormone response element of DNA, selective transcription occurs, resulting in the synthesis of mRNAs. This leads immediately to the production of new proteins and a change in metabolic activity in target cells. Group I ligands include androgens, calcitriol, estrogens, glucocorticoids, mineralocorticoids, progestins, and thyroid hormones.

   Group II hormones are hydrophilic polypeptides or catecholamines that have a short ($t_{\frac{1}{2}}$ = minutes) plasma half-life because they are not bound to transport proteins. Their receptors are located in the plasma membrane and lead to the formation of second messengers such as cyclic AMP (cAMP), the widespread cyclic nucleotide derived from ATP by the enzyme adenylate cyclase.

   The **adenylate cyclase** system is extraordinarily complex. Interaction between ligand and receptor can lead either to the activation or inactivation of adenylate cyclase. For example,

ACTH, ADH, FSH, glucagon, PTH, and TSH among others are all hormones that stimulate adenylate cyclase. Here we will call them H_s and they bind to receptor, R_s. In contrast, acetylcholine, angiotensin II, and somatostatin among others are all hormones that inhibit adenylate cyclase. Here we will call them H_i and they bind to receptor, R_i. This process is mediated by **GTP-dependent regulatory proteins**, called G_s and G_i. These **G proteins** are each composed of three subunits, α, β, and γ. The β and γ subunits are similar in G_s and G_i but the α subunits are distinct (α_s and α_i). This trimer, in the presence of GTP, is converted to a GTP-α_s or GTP-α_i complex and a β-γ complex. The GTP-α_s complex stimulates the catalytic portion of adenylate cyclase, leading to formation of cAMP from ATP. The GTP-α_i complex inhibits the catalytic portion of adenylate cyclase, preventing formation of cAMP from ATP.

In higher (eukaryotic) organisms, cAMP binds to **protein kinase** which consists of an inactive heterotetramer with 2 **regulatory subunits** and 2 **catalytic subunits**. cAMP binds to the regulatory subunits and dissociates them from the catalytic subunits, which are thus rendered enzymatically active. The catalytic subunits in turn catalyze the transfer of the gamma-phosphate from ATP to serine or threonine residues generating many different **phosphoproteins** including **calcium/calmodulin-dependent kinase, myosin light chain kinase, phosphorylase kinase,** and **pyruvate dehydrogenase kinase**. These phosphoproteins are then thought to have diverse physiologic effects. Regulation can occur at several levels including degradation of cAMP by **phosphodiesterase** and/or dephosphorylation by **phosphoprotein phosphatases**.

Cyclic GMP, which is synthesized from GTP under the influence of **guanylate cyclase**, can serve as a second messenger for **atrial natriuretic factor** (ANF). ANF binds to the membrane guanylate cyclase and causes dramatic increases in cGMP levels which in turn leads to diuresis, vasodilation, and inhibition of aldosterone secretion.

Examine the labeled diagram of the neuromuscular junction in **Figure 5.1** and then match the labeled structure with the most appropriate description of its physiological role in the motor end-plate.

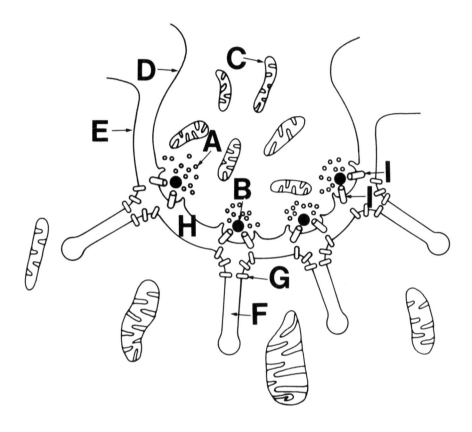

Figure 5.1

416. This intercellular gap between the alpha motor neuron and the myocyte contains receives neurotransmitter released from synaptic vesicles.

417. These membrane delimited structures fuse with the presynaptic membrane, releasing neurotransmitter molecules into the synaptic cleft.

418. When neurotransmitter binds to these structures, a chemically activated channel opens, causing depolarization of the muscle cell.

419. These invaginations of the postsynaptic membrane give rise to an increase in membrane surface area and therefore to an increase in number of acetylcholine receptors.

420. These structures in the presynaptic membrane are thought to cause assembly of groups of neurotransmitter rich vesicles near the presynaptic membrane.

421. Clathrin rich coated pits form in this structure.

422. The basal lamina of the muscle cell binds acetylcholinesterase in this structure.

ANSWERS AND TUTORIAL ON ITEMS 416-422

The answers are: **416-H; 417-A; 418-G; 419-F; 420-B; 421-D; 422-H**. **Figure 5.1** is a diagram of the **neuromuscular junction**, the site of intimate contact between a motor neuron and a skeletal muscle fiber. The **presynaptic** (neuronal) **membrane** (D) has **dense bars** (B) which are thought to be involved in grouping **synaptic vesicles** (A) in clusters. The synaptic vesicles are membrane-bound sacs, synthesized near the Golgi apparatus in the cell body of the motor neuron in the ventral motor horn of the spinal cord. Synaptic vesicles are then transported in an **anterograde** fashion toward the motor end plate. About 300,000 of these vesicles occur in a single motor end-plate. Each vesicle contains about 10,000 molecules of the neurotransmitter, **acetylcholine**.

When the nerve action potential reaches the end-plate, it causes **voltage-gated calcium channels** near the dense bars to open, stimulating synaptic vesicles to fuse with the presynaptic membrane and empty their acetylcholine into the **synaptic cleft** (H). Acetylcholine diffuses rapidly across the synaptic cleft where it binds to the α subunit of the **acetylcholine receptor** (G) concentrated in the upper portions of the **subneural clefts** (junctional folds) (F) in the **postsynaptic membranes** (E).

When no acetylcholine is present, the acetylcholine receptor is in a conformation that prevents passage of small ions through its aqueous central channel. When acetylcholine binds to the α subunit of the acetylcholine receptor, a conformational change occurs in the receptor, opening the aqueous channel to allow the passage of ions. Na+ ions rush into the muscle cell, causing its depolarization and eventual contraction. Negative charge in the lumen of the aqueous channel prevents passage of anions. After release from the acetylcholine receptor, this neurotransmitter is rapidly destroyed by the enzyme **acetylcholinesterase**, bound to the **basal lamina** occupying the synaptic cleft.

Even in the absence of neural stimulation, now and then a single vesicle fuses with the presynaptic membrane, releasing acetylcholine into the synaptic cleft. This results in a **miniature end-plate potential** of about 1 mV with a duration of a few msec. When a stimulatory action potential arrives at the end-plate, 200-300 vesicles suddenly fuse with the presynaptic membrane, resulting a **large end-plate potential** which elicits an **action potential** in the muscle fiber. This depolarization event is followed rapidly by hydrolysis of acetylcholine, recovery of choline in the nerve fiber for reutilization in new acetylcholine synthesis and repolarization of the motor end-plate.

The membranes of synaptic vesicles fuse with and become incorporated into the presynaptic membrane as they release acetylcholine into the synaptic cleft. For continuing function of the nerve cell, the membranes of synaptic vesicles must be recovered from the presynaptic membrane and must be resynthesized into new synaptic vesicles reloaded with acetylcholine. Vesicle membrane is recovered from the presynaptic membrane by **endocytosis**. Immediately after the action potential passes, coated pits form. These have the contractile protein **clathrin** associated with their cytoplasmic faces. Clathrin contraction results in the coated pits invaginating and detaching from the presynaptic membrane, to form new synaptic vesicles that then rapidly become packed with a fresh supply of acetylcholine, readying them for a new cycle of neurotransmitter release.

Items 423-427

The following set of items pertains to the regulation of contractile activity in smooth muscle cells. Choose the best answer.

423. All of the following are true for myosin light chain kinase (MLCK) **EXCEPT**:

    (A)   It catalyses ATP transfer.
    (B)   It is inactivated by phosphatase.
    (C)   It is activated by Ca.
    (D)   It catalyses phosphorylation of myosin light chain.
    (E)   When bound to Ca^{2+}-calmodulin, it becomes inactivated.

424. All of the following are true for cAMP-mediated regulation of smooth muscle contraction **EXCEPT**:

    (A)   occurs when epinephrine stimulates smooth muscle
    (B)   cAMP activates protein kinase
    (C)   when ATP phosphorylates MLCK, it inactivates it
    (D)   when MLC becomes phosphorylated, muscles relax
    (E)   binding to β-adrenergic receptor relaxes smooth muscle

425. All of the following are true for Ca^{2+}-calmodulin regulation of contraction **EXCEPT**:

    (A)   When Ca^{2+} falls, muscle contraction is activated.
    (B)   When Ca^{2+} rises, MLCK is activated.
    (C)   When MLCK is activated, it consumes ATP.
    (D)   When MLCK is activated, it allows myosin binding to actin.
    (E)   When phosphatase dephosphorylates myosin, it cannot bind to actin.

426. Which of the following skeletal muscle proteins is most similar, structurally and functionally, to calmodulin?

(A) actin
(B) myosin
(C) tropomyosin
(D) troponin C
(E) α-actinin

427. At low Ca^{2+} concentrations in the cytosol of smooth muscle (<1 μM), caldesmon binds to which complex to prevent its interaction with myosin, thus causing relation?

(A) troponin C-troponin I
(B) α-actinin-tropomyosin
(C) actin-tropomyosin
(D) tropomyosin-troponin C
(E) troponin T-actin

ANSWERS AND TUTORIAL ON ITEMS 423-427

The answers are: **423-E; 424-D; 425-A; 426-D; 427-C. Smooth muscle cells** contain an actin-myosin based contraction system with many fundamental similarities to skeletal and cardiac muscle but **without** the regular arrangement of thick and thin filaments into **sarcomeres**. The regulation of smooth muscle contraction is also quite different from skeletal and cardiac muscle.

In the relaxed state, smooth muscle cells have myosin molecules in an inactive state where they are unable to polymerize due to the fact that the tail of the myosin rod is associated with **myosin light chains** (MLC). **Myosin light chain kinase** (MLCK) can be activated in a number of different ways. When MLCK is activated, it phosphorylates MLC via ATP. Phosphorylation of the MLC has two effects: (1) it activates an actin binding site in MLC and (2) it prevents the myosin tail from binding to the MLC and instead allows the myosin molecules to bind to one another to form bipolar filament bundles, similar in many respects to thick filaments of striated muscle. These now interact with actin and contraction occurs. This system can be regulated in several different ways.

Ca^{2+}-calmodulin is one important system for regulation of MLCK activity and thus smooth muscle contraction. Calmodulin bears a great sequence homology to **troponin C**, the Ca^{2+}-binding regulatory component of the troponin C complex used to regulate contraction of striated muscle. When intracellular Ca^{2+} rises, it binds to calmodulin to form a complex. This complex then interacts with and **activates MLCK** which **phosphorylates the MLC** to allow contraction. A **phosphatase** activity in cells can **dephosphorylate MLC** and cause **relaxation**.

Epinephrine binds to the β-adrenergic receptors of cells, activating adenylate cyclase and causing the synthesis of cAMP. Smooth muscle cells have a **cAMP-dependent protein kinase** which phosphorylates MLCK at a site near its calmodulin-binding domain. This prevents binding of calmodulin and inactivates the MLCK, causing relaxation.

When the intracellular Ca^{2+} is < 1 micromolar, **caldesmon** interacts with tropomyosin and actin, and thus inhibits muscle contraction because actin-myosin interactions are prohibited. When the intracellular Ca^{2+} rises, the Ca^{2+} binds to **calmodulin**. This complex in turn interacts with caldesmon and decreases its affinity for actin. Subsequently, muscle contraction can occur.

Items 428-430

You encounter a patient who is phenotypically female. Karyotype analysis reveals a normal looking 46, XY karyotype. Molecular biological investigation of the Y chromosome DNA sequences shows a deletion of the testis determining factor SRY.

428.   The testis determining factor encodes for a

(A)   zinc finger protein
(B)   *ras* protooncogene product
(C)   *fos* protooncogene product
(D)   insulin-like growth factor
(E)   zipper protein

429.   If the patient had a 46,XX karyotype with the TDF region of the Y chromosome translocated into the X chromosome the patient would be

(A)   phenotypically female
(B)   phenotypically male
(C)   a hermaphrodite
(D)   a Turner syndrome
(E)   a Klinefelter syndrome

430.   The testis determining factor gene product is a

(A)   translation factor
(B)   repressor
(C)   DNA polymerase inhibitor
(D)   transcription factor
(E)   reverse transcriptase

ANSWERS AND TUTORIAL ON ITEMS 428-430

The answers are: **428-A; 429-B; 430-D**. The testis determining factor **SRY** (for sex reversal Y) is a gene product sharing significant homology with other **zinc finger proteins**, a class of **transcription factors**. Some growth factors and protooncogene products are also transcription factors. When the DNA sequence encoding the testis determining factor is deleted from the Y chromosome, **sex reversal** results. A male karyotype and a female phenotype will occur. Conversely, when the DNA sequence encoding for the testis determining factor is translocated from the Y chromosome to the X chromosome, sex reversal will result again. In this instance, a female karyotype and a male phenotype will be associated. The testis determining factor drives the differentiation of the indifferent gonad in the male direction. Differentiation of a testis then directs sexual differentiation in a male direction. **Turner syndrome** is caused by a monosomy of the X chromosome (45, X). **Klinefelter syndrome** is cause by aneuploidy of sex chromosomes (47, XXY).

Items 431-433

Recent discoveries in developmental genetics reveal a connection between transforming growth factor β (TGF-β), nerve growth factor (NGF), homeobox gene proteins, and the *fos* protooncogene product.

431.   All of these polypeptides

   (A)   are DNA polymerases
   (B)   are translation factors
   (C)   are transcription factors
   (D)   bind to mRNA
   (E)   block tRNA turnover

432.   Their chief mechanism of action is to

   (A)   promote specific mRNA synthesis
   (B)   inhibit DNA synthesis
   (C)   promote specific tRNA synthesis
   (D)   suppress gene expression
   (E)   activate genes for rRNA synthesis

166

433. They are thought to result in differentiation by

    (A)   repression of gene expression
    (B)   controlling differential gene expression
    (C)   promoting turnover of the rough endoplasmic reticulum
    (D)   converting rough endoplasmic reticulum into smooth endoplasmic reticulum
    (E)   stimulation of mitochondrial DNA polymerase

ANSWERS AND TUTORIAL ON ITEMS 431-433

The answers are: **431-C; 432-A; 433-B. TGF-β, NGF, homeodomain proteins**, and the *fos* **gene product** are all examples of **transcription factors**. They bind to DNA causing its uncoiling and promote its transcription into specific mRNAs. The action of these transcription factors is thought to represent the fundamental basis for differential gene expression. All cells in an organism are thought to have the same DNA complement. The formation of the multiple differentiated cell types within an organism, e.g., neurons, muscle cells and liver cells, is due to selective activation of some subset of genes within the precursor of that individual cell type. Once these genes are activated by transcription factors, specific transcription and translation results in the formation of an array of proteins peculiar to that highly differentiated cell type.

Items 434-437

Match the protooncogene in the list of answers below with the most appropriate description in the items below. Answers may be used once, more than once, or not at all.

    (A)   *ras* protooncogene
    (B)   *jun* protooncogene
    (C)   *src* protooncogene
    (D)   *sis* protooncogene
    (E)   *erb B* protooncogene

434.   encodes for platelet derived growth factor

435.   encodes for GTP-binding (G) protein

436. encodes for epidermal growth factor receptor, a tyrosine kinase

437. encodes for a nuclear protein transcription factor

ANSWERS AND TUTORIAL ON ITEMS 434-437

The answers are: **434-D; 435-A; 436-E; 437-B. Protooncogenes** are normal genes involved in the synthesis of proteins essential for regulation of cellular growth and proliferation. **Oncogenes** are mutated forms of these protooncogenes and are expressed in many cancer cells that have lost their normal ability to control growth and proliferation. Many are mutated forms of normal genes and when they become mutated, they can lead to malignant transformation. Genetic rearrangements such as translocation, duplication, selective amplification, or loss of regulatory elements, now lead to the production of abnormal growth controlling substances. For example, the protooncogene *sis* encodes for **platelet derived growth factor** (PDGF). When it becomes mutated, cells produce a large quantity of PDGF-related polypeptide leading to prolonged stimulation of proliferation. The *ras* protooncogene encodes for **G proteins**. Under normal growth control, G proteins bind GTP, are activated and stimulate cell growth, and then cleave GTP to GDP and are turned off again. Mutated *ras* oncogenes bind GTP but do not cleave it and therefore provide a chronic growth stimulatory signal. Another protooncogene, *erb B*, encodes for a transmembrane **epidermal growth factor (EGF) receptor**, a tyrosine kinase. When EGF binds to its receptor, the tyrosine kinase activity of the receptor phosphorylates other cytoplasmic proteins leading to stimulation of growth. Mutated *erb* oncogenes have kinase activity without growth factor stimulation, leading to uncontrolled proliferation typical of cancer cells.

The *jun* protooncogene is one of many that encode for nuclear proteins that function as **transcription factors**. Others include *myc*, *myb*, *fos*, *rel*, and *erb A*. All of these transcription factors bind to promoters and other regulatory elements in the genome, and serve to regulate production of specific mRNAs from particular genes (transcription).

Furthermore, many tumor-causing viruses, especially RNA containing **retroviruses**, have incorporated portions of the genome of infected cells into their own genome. The RNA genome of these retroviruses is copied into double stranded DNA by viral **reverse transcriptase** and then incorporated into the host genome where it is replicated along with other host genes. Viruses then receive mutated host genes which they transfer to infected host cells and cause their transformation.

The following items pertain to the role of homeobox genes in the genetic control of developmental function. Choose the best response.

438.   All of the following are true concerning homeobox genes **EXCEPT**:

    (A)   They encode for transcription factors.
    (B)   They have few highly conserved base sequences.
    (C)   They control limb bud development.
    (D)   They control segmental arrangement of body.
    (E)   Their expression is altered by retinoic acid.

439.   Which teratogenic compound is most well characterized as having effects on homeobox gene expression?

    (A)   retinoic acid
    (B)   ethanol
    (C)   phenytoin
    (D)   diethylstilbestrol
    (E)   warfarin

440.   Which technique would be most directly applicable for examination of regional expression of homeobox genes?

    (A)   immunocytochemistry
    (B)   scanning electron microscopy
    (C)   *in situ* hybridization
    (D)   polymerase chain reaction
    (E)   gel electrophoresis and Western blot analysis

ANSWERS AND TUTORIAL ON ITEMS 438-440

The answers are: **438-B; 439-A; 440-C. Homeobox genes** were first discovered when scientists studied homeotic mutations in *Drosophila*. They found that certain homeotic mutations (i.e., those controlling segment-specific structures) were the result of mutation in homeobox genes. These genes are **highly conserved** across the animal kingdom. Homeobox genes are called **Hox** genes in mammals and a large array of these genes have been described in mice and humans. These genes encode for **transcription factors** that bind directly to DNA and regulate gene expression. Specific mRNAs are produced as a result of Hox gene function. Hox gene expression has been implicated in the control of the development of limb buds (different Hox genes are expressed in different portions of the developing limb bud) and in segmental development of the mammalian body pattern (different Hox genes are expressed at different segmental levels in a cranial to caudal sequence).

The control of Hox gene expression is partially mediated by **retinoic acid**, a potent teratogen thought to promote overexpression of Hox genes and thus limb and craniofacial anomalies. Hox genes have been cloned and radioactive probes have been synthesized, allowing convenient study of regional expression of Hox genes by in situ hybridization in whole embryos or in sectioned embryos.

Items 441-446

Match the specific class of developmental/genetic control mechanism in the answers with the most appropriate description of that class in the items below. Answers may be used once, more than once, or not at all.

(A)   Retinoic acid
(B)   Protooncogenes
(C)   Tumor suppressor gene
(D)   SRY gene
(E)   Hox genes

441.   The region of the male sex chromosome thought to be responsible for testicular differentiation.

442.   These genes regulate segmental morphogenesis in embryos.

443.   These are the wild-type counterpart of genes that control malignant transformation.

444.   This is a known teratogen. In experimental models, it is known to control anterior-posterior specification in limb bud and central nervous system development.

445.   The *kit* gene, encoding for a transmembrane protein with tyrosine kinase activity, is a member of this class.

446.   The retinoblastoma gene *Rb* is a member of this class of genes.

ANSWERS AND TUTORIAL ON ITEMS 441-446

The answers are: **441-D; 442-E; 443-B; 444-A; 445-B; 446-C. Retinoic acid** (A) is a potent teratogen in laboratory animals and its use is contraindicated in pregnant women. In laboratory animals, it has been shown to cause limb defects and abnormalities in craniofacial morphogenesis. It is thought to produce overexpression of homeobox **(Hox) genes** (E) controlling proximal-distal specification in limbs, as well as segmental arrangement in the head, neck, and trunk.

  Protooncogenes (B) are the wild-type forms of oncogenes. They encode for various growth regulatory molecules such as platelet derived growth factor (*sis*), tyrosine kinase growth factor receptors (*kit, ret,* and *erb B*), membrane bound nonreceptor tyrosine kinases (*sec, lok*), G-proteins (*ras*), cytoplasmic protein kinases (*raf*), and nuclear transcription factors (*myc, fos, jun, rel,* and *erb A*).

  Tumor suppressor genes such as the retinoblastoma gene (*Rb*) encode a protein that binds to DNA and prevents mRNA synthesis required to progress through the cell cycle. Mutated *Rb* genes produce aberrant proteins that do not prevent cells from dividing.

  SRY genes (D) are located in a small segment of the short arm of the Y chromosome. It encodes for a **zinc-finger-containing protein** which serves as a transcription factor regulating **testicular differentiation** and thus full male phenotypic expression.

Examine the scanning electron micrograph of a nonconfluent group of fibroblasts *in vitro* below in **Figure 5.2** and then choose the **ONE** best response to the items below.

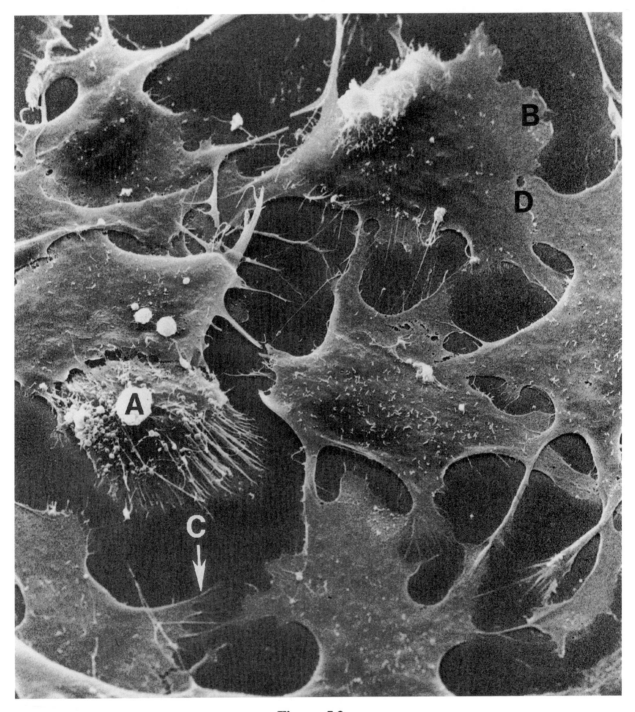

Figure 5.2

447. The cell labeled A is in which portion of the cell cycle?

 (A) S
 (B) G$_1$
 (C) interphase
 (D) M
 (E) G$_0$

448. The structure labeled B is best described as a

 (A) filopodium
 (B) lamellipodium
 (C) pseudopod
 (D) uropod
 (E) retraction fiber

449. The structure labeled C is best described as a

 (A) filopodium
 (B) lamellipodium
 (C) pseudopod
 (D) uropod
 (E) retraction fiber

450. If this culture were treated with colchicine for 24 hours and culture conditions were adequate to support continued cell growth, the relative proportion of cells of the sort labeled A would

 (A) increase
 (B) decrease
 (C) remain unchanged
 (D) insufficient information to answer question

451. Transformation of these fibroblasts with Rous sarcoma virus would have what effect on overall cell culture state?

 (A) cells would die due to cytopathic effects
 (B) cells would become rounded up and contact inhibition of growth control would be lost
 (C) cells would flatten and spread out
 (D) cells would form larger lamellipodia
 (E) cell migration rate would decline due to contact inhibition of locomotion

452. At location D, what phenomenon is most likely to occur?

   (A) filopodia will increase in number explosively
   (B) mitosis
   (C) lamellipodia will contact and their activity will decline
   (D) pinocytosis
   (E) endocytosis

ANSWERS AND TUTORIAL ON ITEMS 447-452

The answers are: **447-D; 448-B; 449-A; 450-A; 451-B; 452-C. Figure 5.2** shows a low density, nonconfluent group of fibroblasts *in vitro*. If the culture conditions are adequate to support growth, these cells will gradually increase in density until they become confluent and stop growing. The cell labeled A has rounded up because it is dividing. It is currently in metaphase. Colchicine treatment for 24 hours would dramatically increase the proportion of cells arrested in M. Rous sarcoma virus would transform these fibroblasts, causing them to loose their firm intercellular adhesions. Transformed cells become rounded up and loose contact inhibition of growth control.

Fibroblasts in nonconfluent cultures are flattened and spread on the substratum because they are firmly attached and are spread under tension by the expansive activity of lamellipodia (B) and filopodia (C), the locomotory organelles of fibroblasts and other cells *in vitro* and *in vivo*. When moving lamellipodia contact and adhere (at D), their locomotory activity ceases and cells will not continue to move in the direction of contact between dominant lamellipodia. This phenomenon is known as contact inhibition of locomotion.

Items 453-455

The following items pertain to basic functional mechanisms of adrenergic receptors. Choose the **ONE** best response.

453. All of the following are true concerning α_1 adrenoreceptors **EXCEPT**:

   (A) agonist binding increases adenylate cyclase activity
   (B) agonist binding activates G-protein
   (C) agonist binding activates phospholipase C
   (D) agonist binding causes an increase in intracellular inositol triphosphate (IP_3)
   (E) agonist binding stimulates Ca^{2+}-dependent protein kinase

174

454. All of the following are true concerning α_2 adrenoreceptors **EXCEPT**:

   (A)   they activate inhibitory G-protein
   (B)   no increase in phospholipase C activity occurs
   (C)   cause platelet aggregation
   (D)   cAMP levels increase
   (E)   no increase in IP_3 is observed

455. All of the following are true concerning β adrenoreceptors **EXCEPT**:

   (A)   they activate stimulatory G-protein
   (B)   cAMP levels increase
   (C)   protein kinase levels cause enzyme phosphorylation
   (D)   intracellular ATP is increased
   (E)   phosphorylated enzymes produce biological effect

ANSWERS AND TUTORIAL ON ITEMS 453-455

The answers are: **453-A; 454-D; 455-D.** These three different kinds of **adrenoreceptors** are activated and have their biological effects promoted by three fundamentally different mechanisms. When an **agonist** (e.g., catecholamine) binds to the α_1 **receptor**, it activates a **G-protein** which then activates **phospholipase C**, an enzyme which releases **inositol triphosphate** (IP_3) and **diacylglycerol** (DAG) from membrane phosphinositides. IP_3 causes a transient release of stored Ca^{2+}, leading to activation of Ca^{2+}-**dependent protein kinase** and the biological effect. Rising DAG levels will also activate **protein kinase C**. Some of the biological effects of α_1 receptor stimulation include vascular smooth muscle contraction, pupillary dilation, glycogenolysis, and increased force of cardiac muscle contraction.

The β and α_2 receptors have their actions through changing cAMP levels. When an agonist binds to the β **receptor**, it triggers the **stimulatory G-protein** which stimulates **adenylate cyclase** and leads to conversion of ATP to **cAMP**. The increased cAMP then causes **protein kinase** to become activated, leading to enzyme **phosphorylation** and biological effect. Biological effects of agonist binding to β receptors include increased force and rate of cardiac muscle contraction, smooth muscle relaxation, activation of glycogenolysis, and activation of lipolysis.

When an agonist binds to the α_2 **receptor**, it triggers the **inhibitory G-protein**, blocking adenylate cyclase activity and causing a **decrease** in cAMP levels. Stimulation of α_2 receptors causes **platelet aggregation** but it is not clear if this effect is due directly to decreased cAMP levels or to some other effect. It also causes inhibition of lipolysis.

Items 456-461 refer to the mechanisms by which extracellular signals are transduced into cellular response mechanisms. For each item select the **ONE** lettered answer that **BEST** corresponds.

456. Which of the following signalling compounds does **NOT** enter cells by passive diffusion through the plasma membrane?

   (A)   estrogen
   (B)   thyroxine
   (C)   acetylcholine
   (D)   vitamin D
   (E)   retinaldehyde

457. Membrane permeant signalling ligands

   (A)   are often carried to cells by transport proteins that keep the compounds from precipitating out of solution in the bloodstream.
   (B)   trigger rapid responses that involve little if any gene activation.
   (C)   can only reach their intracellular targets by crossing many membranes in addition to the plasma membrane.
   (D)   bind very weakly to their receptors.
   (E)   are frequently small polypeptides.

458. Select the **FALSE** completion. Nitric oxide

   (A)   diffuses freely across the plasma membrane of target cells.
   (B)   is produced by deamination of arginine residues.
   (C)   is short acting because it is rapidly degraded in most cells.
   (D)   causes smooth muscle cells in the walls of blood vessels to relax.
   (E)   usually exerts its effects by inhibiting guanylyl cyclase and thus reducing intracellular levels of cyclic GMP.

459. The intracellular effects of acetylcholine

   (A)   require that cytosolic concentrations of the compound reach millimolar levels.
   (B)   depend on its ability to trigger transcription of DNA into RNA.
   (C)   are quickly blocked by inhibitors of protein synthesis.
   (D)   result from its binding to ion channels and the opening of such channels.
   (E)   can only be reversed slowly because it is stabilized by binding proteins in the cytoplasm.

460. Trimeric GTP binding proteins

    (A)   are members of the "seven-pass" transmembrane family of proteins.

    (B)   form by the association of three identical polypeptides.

    (C)   become activated by the binding of GTP to a receptor site on the extracellular surface of the plasma membrane.

    (D)   include such proteins as myosin and rhodopsin.

    (E)   work by a complex pathway that initially activates a set of membrane bound protein phosphatases.

461. Which of the following is **LEAST** likely to occur in the transduction pathway leading from binding of a signal compound to a seven-pass transmembrane protein to activation of protein kinase A?

    (A)   Binding of the α subunit of a trimeric G-protein to the cytoplasmic portion of the transmembrane receptor.

    (B)   Release of GDP from the α subunit.

    (C)   Release of the α subunit from the β and γ subunits of the trimeric G-protein.

    (D)   Binding of GTP by the α subunit.

    (E)   Transfer of the GTP from the α subunit to (and activation of) nearby membrane linked adenylate cyclase.

ANSWERS AND TUTORIAL ON ITEMS 456-461

The answers are: **456-C; 457-A; 458-E; 459-D; 460-A; 461-E.** Eucaryotic cells respond to signals from the exterior by a variety of mechanisms. It has been found very useful to distinguish pathways of **signal transduction** that involve receptors within the plasma membrane from those that do not involve membrane-bound receptors. Many signal transduction pathways involve signalling ligands that diffuse freely through the plasma membrane and bind to intracellular receptors; such ligands include steroids (e.g., estrogen) and a variety of lipid soluble compounds such as thyroxine, vitamins, and retinoids. These enter cells freely and then interact with one or more receptors through which their effects are mediated. In contrast, many compounds such as acetylcholine do not freely enter cells and instead exert their effects by binding to **transmembrane proteins** that function as **signal receptors** that are the first element in a signal transduction pathway. Ligands that exert their effects by diffusing freely into target cells are a diverse group, but they share a number of features. Being lipid soluble and hydrophobic they have low solubility in blood and are usually maintained in the circulation by binding to carrier proteins. They are released slowly from such carriers and, upon entering target cells they bind slowly to their receptors and are only slowly released from their receptors. The net result is that such compounds act slowly and for prolonged periods of time; many such signalling compounds act to trigger gene transcription and thus protein synthesis. One exception to this general scheme is **nitric oxide** (NO). NO is produced in many cells by the deamination of arginine and can enter neighboring cells rapidly by diffusion. In smooth muscle cells it activates guanylate cyclase causing elevation of levels of cGMP which leads to relaxation of the muscle. Because the NO is rapidly (less than 1 minute) degraded, its actions are turned off quite rapidly. In contrast to mechanisms involving signal molecules that diffuse freely and rapidly across the plasma membrane, there are a variety of signal transduction pathways involving poorly permeant (usually hydrophilic) compounds that bind to membrane proteins which transduce the extracellular signal to the cell interior. One such class of signalling compounds are those, like acetylcholine that bind to transmembrane proteins that make up **ion channels**. Such transduction pathways cause rapid influx (or efflux) of specific ions (such as Ca^{2+}), which then trigger specific response pathways. Such signal compounds need not enter the target cell; they are frequently rapidly released from the receptor, and/or rapidly degraded by extracellular enzymes (e.g., cholinesterase) so that their effects are short lived.

One of the most important classes of signal transduction pathways that rely on binding to cell surface receptors is the class that involves trimeric **GTP binding proteins**. These include a family of so-called "seven-pass" transmembrane proteins, meaning that one component of the receptor complex is a protein that has a ligand binding site on the extracellular side of the plasma membrane, snakes back and forth across the plasma membrane seven times, and ends on the cytoplasmic side of the membrane with an effector site. There are a variety of pathways by which binding of the ligand to the extracellular receptor portion leads to activation of a specific cytoplasmic response; one of the best studied is the pathway leading to increased levels of the second messenger cAMP. In this pathway, binding of the stiumlating ligand to the cell surface receptor domain leads to a change in the cytoplasmic part of the seven-pass transmembrane such that it can bind the α subunit of a nearby trimeric G-protein. This binding reduces the affinity of the α subunit for GDP (which is released) and faciltates the binding of GTP. The α subunit then separates from the other two subunits (β and γ) which are inhibitory. The α subunit, freed of the inhibitory portions of the trimer, then binds to nearby adenylate cyclase activating it and leading to increased production of cAMP. The α subunit also hydrolyses its bound GTP and, unless local levels of GTP are high, the α subunit with bound GDP is recaptured by the inhibitory β and γ subunits, those shutting down the cAMP synthesis pathway.

CHAPTER VI
CELL BIOLOGY AND DISEASE STATES

Items 462-465

A 5 year-old boy of northern European ethnic background develops severe anemia, splenomegaly, and jaundice. A light micrograph of his peripheral blood is shown in **Figure 6.1** below.

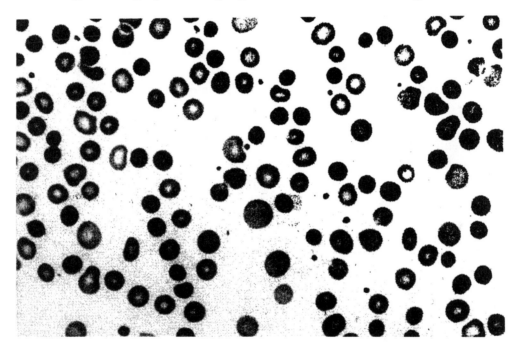

Figure 6.1

462. The most likely diagnosis of this condition is

    (A)   sickle cell anemia
    (B)   β-thalassemia
    (C)   α-thalassemia
    (D)   hereditary spherocytosis
    (E)   malaria

463. Which erythrocyte membrane constituent is abnormal in this disease?

    (A)   peripheral membrane protein
    (B)   integral membrane protein
    (C)   membrane phospholipids
    (D)   membrane glycolipids
    (E)   membrane cholesterol

464. Which protein is affected by this mutation?

    (A)   actin
    (B)   spectrin
    (C)   desmin
    (D)   integrin
    (E)   tubulin

465. In the most common form of this disease, which interaction is defective as a result of the mutation?

    (A)   F-actin formation
    (B)   spectrin-lipid interaction
    (C)   ion transport proteins loose their hydrophobic domains and can't become integral transmembrane proteins
    (D) α and β tubulin fail to associate
    (E)   actin-talin interaction

ANSWERS AND TUTORIAL ON ITEMS 462-465

The answers are: **462-D; 463-A; 464-B; 465-B**. **Figure 6.1** shows a blood film of a patient with **hereditary spherocytosis** (HS). In its most common form, this disease is caused by an autosomal dominant mutation leading to spherical erythrocytes with abnormal fragility. Thus, it is a variety of hemolytic anemia. As a result of mutations in several erythrocyte proteins including ankyrin, α and β spectrin, and protein 4.2, there is a loss of the protein **spectrin**, and concomitant defects in interaction between spectrin and **membrane lipids**. This results in a loss of redundant erythrocyte surface area, increased erythrocyte fragility, and hemolytic anemia.

A 7 year-old African-American girl is diagnosed with a kind of hemolytic anemia. **Figure 6.2** below shows a scanning electron micrograph (upper) and a transmission electron micrograph (lower) of her erythrocytes.

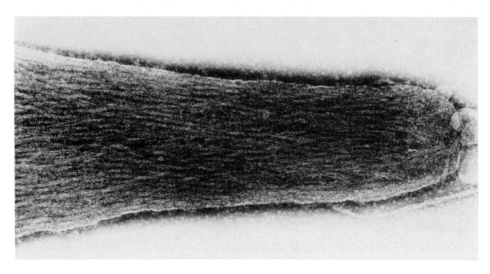

Figure 6.2

466. The most likely diagnosis of her condition is

(A)   hereditary spherocytosis
(B)   hereditary elliptocytosis
(C)   sickle cell anemia
(D) α-thalassemia
(E) β-thalassemia

467. The fibrillar structures in the lower micrograph contain predominantly

(A)   spectrin
(B)   hemoglobin
(C)   actin
(D)   glucose-6-phosphate dehydrogenase
(E)   glycophorin

468. The gene defect in this disease has what primary effect on Hb?

(A)   increases catabolism
(B)   increases solubility
(C)   increases dissociation
(D)   decreases solubility, especially of deoxyHb
(E)   decreases catabolism

469. All of the following physiologic states would favor formation of fibrillar structures in her erythrocytes **EXCEPT**:

(A)   increase in Hb concentration
(B)   decrease in serum PO_2
(C)   increase in concentration of deoxyHb
(D)   respiratory acidosis
(E)   hyperventilation

ANSWERS AND TUTORIAL ON ITEMS 466-469

The answers are: **466-C; 467-B; 468-D; 469-E. Figure 6.2** shows erythrocytes as they appear in **sickle cell anemia**. This disease is caused by a substitution of valine for glutamic acid in the β chain of hemoglobin. Under normal circumstances, the hemoglobin concentration in erythrocytes is extremely high. The point mutation for sickling decreases the solubility of HbS, especially when deoxygenated, leading to the formation of semi-rigid fibrillar aggregates in cells

182

(**Figure 6.2**, lower) and significant deformation of the normally biconcave discs of erythrocytes. These abnormal erythrocyte shapes and excessive rigidity lead to hemolytic anemia.

Items 470-473

The mother of a three year-old boy brings her child to the family doctor concerned because the child was slow to begin walking, has difficulty in walking and standing and fatigues easily. She recalls that the youngest brother of her mother had similar problems and died in his late teens. Tests of the child reveal creatinuria and elevated serum levels of creatine phosphokinase (CPK).

470.   Which of the following diagnoses is most likely to be correct?

     (A)   nemaline myopathy
     (B)   poliomyelitis
     (C)   Duchenne's muscular dystrophy
     (D)   myasthenia gravis
     (E)   mitochondrial myopathy

471.   Enzymatic assay of a punch biopsy from the gluteus maximus is **MOST** likely to show

     (A)   unusually high levels of CPK
     (B)   elevated levels of myoglobin
     (C)   normal amounts of myosin
     (D)   reduced activity of lactate dehydrogenase
     (E)   reduced activity of lysosomal hydrolases

472.   A principal factor in the occurrence of this disease is

     (A)   reduced amount of the protein dystrophin
     (B)   inability of actin filaments to bind to the Z-line
     (C)   overproduction of a myosin isoform with reduced ATPase activity
     (D)   absence of gap junctions between muscle fibers
     (E)   tropomyosin is not produced, hence thick filaments are unstable

473.   All of the following are normally properties of the protein that is altered in this disease **EXCEPT**:

     (A)   great sequence homology with spectrin
     (B)   slight sequence homology with calmodulin
     (C)   hydrolysis of ATP
     (D)   association with the sarcolemma
     (E)   binding to actin filaments

ANSWERS AND TUTORIAL ON ITEMS 470-473

The answers are: **470-C; 471-D; 472-A; 473-C**. This child is suffering from **Duchenne's muscular dystrophy** (DMD). This genetically linked condition is maternally transmitted, only to male children. It usually manifests in the first decade of life as difficulty in standing and walking. Muscle weakness is initially manifested in muscles of the hip girdle and upper legs. Later the condition involves muscles of the arms and respiratory system, with death usually occurring before age 20. Biopsies show muscle fibers of highly variable size, as well as degenerating fibers and fibers being broken down and phagocytosed. Patients fail to produce the protein **dystrophin**. This is a member of the spectrin superfamily, with a MW = 427,000. Dystrophin is found underneath the sarcolemma in normal muscle, especially where membrane folds are formed, e.g., at myotendinous and neuromuscular junctions. It also interacts with actin filaments, stabilizing their side-side association as well as binding to the amino-terminus. Muscle from DMD patients shows a tendency to fragility and rupture. They also show reduced levels of enzymes normally found in the sarcoplasm with concomitant increases in serum levels of creatine phosphokinase, lactate dehydrogenase and glucose phosphoisomerase. Urinary levels of creatine are usually elevated. It is believed that one of the normal functions of dystrophin is to stabilize the sarcolemma against the physical stresses caused by frequent folding during cycles of contraction and relaxation.

ITEMS 474-476

An 18 year-old African-American from a small town in Alabama is seen at a rural clinic because of a broken radius, apparently suffered during a friendly bout of arm wrestling with his older sister. Physical examination reveals several poorly healed fractures. The young man confirms their occurrence and comments that a 40 year-old uncle has similar problems related to "clumsiness". Also, the youth has reduced hearing and his sclera are somewhat blue.

474.   This patient is suffering from

    (A)   scurvy
    (B)   battered child syndrome
    (C)   chondromalacia
    (D)   osteogenesis imperfecta
    (E)   Marfan syndrome

475. The most **relevant** finding upon analysis of a sample of bony tissue would likely be

   (A)   reduced levels of collagen type I in the extracellular matrix
   (B)   Haversian canals filed with crystals of hydroxyapatite
   (C)   numerous deposits of lead sulfate
   (D)   unusually high concentrations of elastin
   (E)   elevated concentrations of fibronectin

476. Testing of the uncle reveals the same defect as the patient. This is consistent with the diagnosis because

   (A)   the syndrome is linked to the Y chromosome
   (B)   such disorders are autosomal dominant
   (C)   the defect is usually found in males who play many sports
   (D)   impoverished families often have dietary deficiencies
   (E)   the condition only appears long after puberty

ANSWERS AND TUTORIAL ON ITEMS 474-476

The answers are: **474-D; 475-A; 476-B**. The patient has **osteogenesis imperfecta** (type I). This is an inherited defect (**autosomal dominant**) that leads to reduced production of **type I collagen**. The collagen present in bone matrix is of type I and is produced by osteoblasts. The mechanism of collagen synthesis (by ribosomes of the rough ER), processing, and secretion is closely similar to that occurring in fibroblasts. Individuals with osteogenesis imperfecta have a deletion in one of the genes for the $\alpha(I)$ chain of collagen (I). The deletion leads to production of a shortened $\alpha(I)$ chain. The shortened $\alpha(I)$ chains associate with normal $\alpha 1(I)$ and $\alpha 2(I)$ chains causing formation of defective triple helices that are especially sensitive to proteolytic degradation. Victims have numerous fractures and resulting bone deformities. Other associated effects are dental problems, hearing loss and a bluish tint to the sclera.

Items 477-481

A 3 year-old girl shows a white reflection from the surface of the retina upon ophthalmologic examination. Biopsy of the lesion reveals retinoblastoma.

477. If this disease was of the heritable form, she would also be at increased risk for all of the following diseases **EXCEPT**:

   (A)   lung carcinoma
   (B)   breast carcinoma
   (C)   urinary bladder carcinoma
   (D)   leukemia

478. If this disease was of the sporadic form, all of the following would be true **EXCEPT**:

   (A)   Tumors would be bilateral
   (B)   Later onset more likely
   (C)   Due to homozygous somatic mutation
   (D)   Occurs in majority of retinoblastoma cases

479. The gene responsible for retinoblastoma (*Rb*) is located where in the genome

   (A)   band 13p13
   (B)   band 12q6
   (C)   band 13q14
   (D)   band 12q13
   (E)   band 12p13

480. The *Rb* gene is which type of gene

   (A)   protooncogene
   (B)   oncogene
   (C)   tumor suppressor gene
   (D)   homeobox gene

481. Phosphorylated *Rb* gene product has which significant characteristic?

   (A)   sequesters gene regulatory proteins
   (B)   releases gene regulatory proteins
   (C)   is degraded
   (D)   is allosterically inactivated
   (E)   is a tumor promotor

ANSWERS AND TUTORIAL ON ITEMS 477-481

The answers are: **477-D; 478-A; 479-C; 480-C; 481-D. Retinoblastoma** is a rare malignancy of the retina seen often in young children. In the **hereditary form**, usually both eyes are afflicted by multiple tumors. In addition, lung, breast, and urinary bladder carcinomas are more frequent. Some cases of retinoblastoma are due to a deletion of chromosomal band 13q14. In the more common **sporadic form**, only one eye will be afflicted with a tumor.

The *Rb* gene is a **tumor suppressor gene** and its product is a tumor suppressor protein which can be phosphorylated. In the phosphorylated state, it *Rb* protein **sequesters** regulatory proteins required for cell division. In the dephosphorylated state, *Rb* protein releases the gene regulatory proteins, allowing mitosis to occur. In the absence of the *Rb* gene product, therefore, there is a generalized loss of growth control and appearance of tumors.

Items 482-487

For each numbered item from 482 through 487, choose the lettered compound that most closely corresponds.

    (A)   Botulinum toxin
    (B)   Chloramphenicol
    (C)   Vinblastine
    (D)   Phalloidin
    (E)   Colchicine
    (F)   Puromycin
    (G)   Taxol
    (H)   Griseofulvin
    (I)   Acridine orange
    (J)   Okadaic acid

482. This material is the active agent in the poisonous mushroom commonly referred to as "death's angel". The substance is hepatotoxic, apparently because it binds to actin and interferes with the functioning of microfilaments near bile canaliculi.

483. Although a very efficient inhibitor of bacterial protein synthesis, this compound is no longer used clinically as an antibiotic because it also inhibits mitochondrial protein synthesis.

484. This bacterial product is a potent inhibitor of neuromuscular transmission and has found recent clinical application in the treatment of many disorders, such as spasmodic torticollis, that are due to idiopathic hyperactivity of certain skeletal muscle fibers.

485. An anticancer agent, this compound inhibits proper formation of microtubules required for mitosis, apparently by causing tubulins to form paracrystalline arrays.

486. Extracted from the bark of the yew tree, this compound "locks" tubulin monomers into the polymeric state (preventing microtubule disassembly) and has recently found use as an anti-tumor agent.

487. This compound is an inhibitor of protein phosphatases and inhibits a range of regulatory processes based on control of the level(s) of phosphorylation of specific protein(s).

ANSWERS AND TUTORIAL ON ITEMS 482-487

The answers are: **482-D; 483-B; 484-A; 485-C; 486-G; 487-J. Botulinum toxin** (A) is the product of the bacterium *Clostridium botulinum*; it inhibits skeletal muscle contraction by blocking cholinergic neuromuscular transmission and has been used to ameliorate several symptoms, such as spasmodic torticollis, in which muscle fibers are hyperactive and unresponsive to "normal" control mechanisms.

Chloramphenicol (B) is an antibiotic that inhibits protein synthetic activity of procaryotic ribosomes as well as mitochondrial ribosomes, thus making it unsuitable for long term treatment of bacterial infections.

Vinblastine (C) is a plant alkaloid that binds to tubulin, inducing a variety of paracrystalline aggregates and thus inhibiting cellular functions (such as the mitotic proliferation of tumor cells) that require assembly and disassembly of microtubules.

Phalloidin (D) is the toxic product of the death's angel mushroom, *Amanita phalloides*; it binds tightly to actin microfilaments, such as those around bile canaliculi, hyperstabilizing them and blocking processes that require both assembly **and** disassembly of microfilaments.

The plant alkaloid **colchicine** (E) binds to tubulin, preventing its assembly to form microtubules; it has been used to block mitosis by blocking formation of the mitotic spindle.

Puromycin (F) blocks protein synthesis by both pro- and eucaryotic ribosomes because it mimics an activated amino-acyl-tRNA and induces premature termination of the growing polypeptide chain.

The compound **taxol** (G), derived from the bark of the yew tree, has found increasing use as an anti-cancer agent; it hyperstabilizes microtubules, blocking the disassembly of these tubulin polymers and depleting cells of the pool of tubulin monomers required to form new microtubules at different times and places.

Griseofulvin (H) is also an inhibitor of the polymerization of certain tubulins and thus disrupts such cellular processes as mitosis.

Acridine orange (I) binds to nucleic acids and, at high enough doses, disrupts DNA and RNA synthesis.

Okadaic acid (J), a polyether fatty acid isolated from marine sponges, is one of a growing group of compounds found to block the activity of phosphatases that specifically reverse the phosphorylation of proteins, such as calmodulin, myosin light chains, etc., whose function is regulated by the balance between protein kinase and protein phosphatase activity.

Items 488-497

The following set of items pertain to the genetic basis for cancer. Match the correct answer with the appropriate description in the items below the answers.

(A)   Retinoblastoma
(B)   Wilms' tumor
(C)   Chronic myelogenous leukemias (CML)
(D)   *ras* oncogenes
(E)   *src* oncogenes
(F)   *erb B* oncogenes
(G)   *fos*
(H)   *myc*
(I)   *sis*
(J)   *abl*

488.   Human oncogenes that are detected by DNA transfections.

489.   Deletion of band q14 on chromosome 13 leads to the loss of both copies of the *Rb* gene.

490.   Deletion of band p13 of chromosome 11.

491.   Philadelphia chromosome, derived from chromosome 22 by a reciprocal translocation from chromosome 9.

492.   Nuclear transcription factor.

493.   Nuclear protein, strictly regulated with respect to the cell cycle.

494.   Plasma membrane associated tyrosine-specific protein kinase.

495.   Platelet derived growth factor β subunit.

496. Originally isolated from a chicken sarcoma, a tyrosine-specific protein kinase.

497. A transmembrane epidermal growth factor which is a class I oncogene.

ANSWERS AND TUTORIAL ON ITEMS 488-497

The answers are: **488-D; 489-A; 490-B; 491-C; 492-G; 493-H; 494-J; 495-I; 496-E; 497-F**. Several human malignancies have been linked to chromosomal abnormalities that include deletions or translocations of fragments of chromosomes.

   Retinoblastoma (A) (RB), an ocular tumor of children, is a highly penetrant autosomal dominant trait. The tumors result from loss of function of both *Rb* genes. In some cases, the lesion involves chromosome 13 band q14. Tumor cells are homozygous for the deleted version of chromosome 13. Normal cells from individuals with a high predisposition to the disease can carry one normal chromosome 13 and one with a deletion of band q14, while their tumor cells are homozygous for the abnormal chromosome. The clinical effect of *Rb* gene deletion is through the removal of a *Rb* gene product (pRB 105 is a cell growth suppressor). Functional neutralization of the protective pRB gene product by viral oncoproteins is a common mechanism of cell transformation. Therefore, *Rb* gene, and several other analogous genes are called **tumor suppressor genes**, since it is loss of their function that causes cell transformation. Consequently, mutations at the Rb locus can predispose individuals to other forms of cancer as well as retinoblastoma. Preclinical screening of *Rb* and other tumor-suppressor gene mutations is already proving to be useful resource for understanding human cancer.

   Wilms' tumor (B) is an early childhood tumor of the kidney. The hereditary form of this tumor suggests that the mechanism of induction may involve functional inactivation of **tumor-suppressor genes**. The development of the tumor involves inactivation of both copies of the WT locus on chromosome 11 band p13. Wilms' tumor cells appear to be defective in a chromosome 11 band p13 gene that functions as a suppressor of neoplastic growth of kidney cells. In this respect, the mechanism of tumor induction in WT appears superficially similar to RB.

   The chromosomal abnormality associated with **chronic myelogenous leukemia** (CML) (C), called the **Philadelphia chromosome** (Ph¹), is a small chromosome associated with 90% of CML cases, an invariably fatal cancer involving uncontrolled multiplication of myeloid stem cells. Ph¹ is derived by a reciprocal translocation involving chromosomes 22 and 9. Formation of the Ph¹ leads to the activation of the *abl* **protooncogene** (J) which lies in a portion of chromosome 9 that is translocated to chromosome 22. The *abl* oncogene was first described as the transforming gene of the highly oncogenic Abelson murine leukemia virus, (*v-abl*). The activated *abl* gene product in CML shows significant homology to the **tyrosine kinase family** of oncogenes. Formation of Ph¹ by translocation results in the expression of *abl* as a fusion protein, and it is the hybrid form of *abl* protein that specifically acquires the ability to transform myeloid stem cells.

Human _RAS_ oncogenes (D) are homologous to the mouse _ras_ genes, of the Harvey _sar_coma virus, a highly oncogenic rodent retrovirus. These are class II oncogenes, i.e., they encode plasma membrane GTP-binding (G-) proteins. They were first isolated from human bladder tumor cells. Other _ras_ genes have been isolated from human lung, colon and bladder. The transforming _ras_ genes have been isolated by DNA transfection of fibroblasts in tissue culture and by isolating the foci of rapidly growing transformed cells. Upon subsequent DNA sequence analysis, these human genes turned out to be homologs of mouse oncogenes. Human _RAS_ genes encode a 189 amino acid protein, called p21_ras_, which is associated with the inner surface of the plasma membrane. These proteins are guanine nucleotide binding proteins, which in their activated (oncogenic) form are deficient in GTPase activity. The structural similarity of p21_ras_ to the guanine nucleotide binding proteins (G-proteins), led to an understanding of the mechanism of malignant cell transformation induced by activated _ras_. Mutations either reduce or abolish the GTPase activity of **G-proteins** leading to a constituitively activated p21_ras_, which in turn induces malignant cell transformation. In an experimental system, mammary tumors induced in rats by the carcinogen N-nitroso-N-methylurea have been shown to involve _ras_ gene mutation.

The _src_ **oncogenes** (E) and _erb B_ **oncogenes** are both class I oncogenes, i.e., they encode for a mutation in **protein kinases**. The _src_ gene product is a tyrosine-specific protein kinase located in the plasma membrane. The _erb B_ gene product is the **epidermal growth factor receptor**. It is a transmembrane protein.

The _fos_ (G) and _myc_ (H) **oncogenes** are class IV oncogenes. They encode for **nuclear proteins**. The _sis_ **oncogene** is a class III oncogene, i.e., one encoding for **growth factors**. The gene products of class III oncogenes are secreted from cells. The _sis_ oncogene product is derived from the gene that is responsible for the synthesis of **platelet derived growth factor** (PDGF).

Items 498-503

For each numbered statement regarding a pathological condition, choose the lettered bacterium most likely involved in the disease state.

> (A) *Escherichia coli*
> (B) *Helicobacter pylori*
> (C) *Listeria monocytogenes*
> (D) *Vibrio cholerae*
> (E) *Campylobacter jejuni*
> (F) *Clostridium tetani*
> (G) *Pseudomonas aeruginosa*
> (H) *Clostridium botulinum*

498. The ability of this infectious bacterium to spread from one infected cell to its neighbors is dependent on the ability of the bacterium to nucleate the assembly of actin microfilaments at one end of the bacterium.

499. Although the exact mechanism by which this bacterium causes gastrointestinal ulcers is still in some dispute, its eradication by intensive drug therapy has become widely accepted as the method of choice for treatment of more than 50% of all such ulcerous conditions.

500. The toxic product of this organism blocks neuromuscular transmission, thus limiting the ability of affected muscles to contract.

501. Unrestricted growth of this anaerobic bacterium leads to production and release of a toxin whose inhibitory effects on neuronal and neuro-muscular transmission lead to continuous and uncontrolled muscle spasms.

502. This pathogen is taken up into cells by phagocytosis, after which it secretes a lytic agent that destroys the phagolysosomal membrane and allows entrance of the pathogen into the cell cytoplasm.

503. A motile gram negative bacterium, this organism produces an enterotoxin that binds to the membrane of intestinal epithelial cells stimulating adenylate cyclase and a large increase in intracellular levels of cAMP. Affected epithelial cells secrete an isotonic fluid into the lumen of the intestines causing massive diarrhea with minimal observable damage to the mucosal lining.

ANSWERS AND TUTORIAL ON ITEMS 498-503

The answers are: **498-C; 499-B; 500-H; 501-F; 502-C; 503-D**. Many of the bacteria that cause medical problems are important not just because of the need to treat their effects but also because understanding their mechanisms of action sheds light on basic cell biological processes. *Escherichia coli* (A) is normally resident in the human gastrointestinal tract, but its introduction into other locations in the body can cause massive and uncontrolled bacterial growth. Early attempts to treat such infections with antibiotics like chloramphenicol lead to the realization that this protein synthesis inhibitor was active on mitochondrial ribosomes, as well as those of bacteria. The preferred treatment is to use inhibitors such as erythromycin, that are more specific for bacterial ribosomes.

Until relatively recently it was believed that peptic ulcers resulted primarily from over-secretion of HCl by gastric parietal cells. In a large percentage of patients with gastroenteritis, the mucosa of the pyloric stomach is found to be heavily infected with a spiral bacterial cell - *Helicobacter pylori* (B). Eradication of *H. pylori* with a variety of antibiotics, in combination with bismuth, is an effective treatment of such conditions and greatly reduces the likelihood that the enteritis will lead to gastric ulcers and/or tumors.

Listeria monocytogenes (C), is a pathogenic bacteria that causes e.g., meningitis in newborns; the most common source is from the mother. Taken up by phagocytosis, it produces a lytic agent (listeriolysin) that dissolves the membrane of the vacuole, releasing the bacterium into the cytoplasm, where it proliferates. The bacterial cell wall contains proteins, including one called ActA, which nucleate assembly of actin filaments. The filaments, and the bundles they form, are made preferentially at one end of the bacterium and the polarized assembly of the actin filaments at this end drives the bacteria in the opposite direction through the cytoplasm of the affected cell. This mechanism appears to guarantee frequent contact of bacteria with the host cell plasma membrane; some of these contacts cause cytoplasmic extensions that can press agent adjacent cells and trigger phagocytic uptake of the bacteria and secondary infection of the neighboring cell.

Vibrio cholerae (D) is a motile gram-negative bacterium whose enterotoxin - a protein of MW = 84 Kd - causes cholera. The toxin causes intestinal epithelial cells to overproduce cAMP; the resulting activation of membrane bound ion pumps leads to excessive secretion of isotonic fluid into the lumen of the small intestine, with only minimal finding of damaged cells in the voluminous diarrheal stool.

Campylobacter jejuni (E) attacks mucosal epithelial cells of the small and large intestines, causing damage and loss of the lining epithelium. The resulting diarrhea (reflecting failure of the damaged mucosa to absorb sufficient fluid from the lumen) is often bloody. *C. jejuni* is usually a minor population in the G.I. flora and does not grow well under routine lab test conditions; the importance of this bacterium in causing diarrhea was only realized when culture conditions were modified (incubation at 42° C) to favor its growth.

Clostridium tetani (F) is one of many bacterial strains that exert their pathology by production and release of specific **toxins**. The tetanus toxin is a protein with a dimer MW ≈ 145,000; it is produced when bacterial spores introduced by puncture wounds (the bacteria are non-invasive) germinate under anaerobic conditions (spores do not germinate under conditions of normal tissue oxygen tension) yielding bacterial growth and toxin release. Tetanus toxin has inhibitory affects on neuronal transmission in both the central and peripheral nervous systems; it also can act directly on skeletal muscle fibers, causing contraction even in the absence of nerve stimulation. One of the most characteristic effects is to cause prolonged contractions (tetanus; normal feedback regulatory pathways are bypassed) of skeletal muscle fibers.

Pseudomonas aeruginosa (G) is an opportunistic bacterium that can cause a variety of infections in humans whose defense systems are compromised. Infections frequently occur in hospital settings such as surgical suites; the organism is often conveyed to patients via contaminated catheters or urine receptacles or by healthy staff who carry the bacterium on their skin.

Clostridium botulinum (H) is an anaerobe whose spores are transmitted to humans in foods that are not processed and/or cooked sufficiently. The bacteria release a toxin - botulinum toxin - that causes progressive descending paralysis of skeletal muscle. All variants of the toxin (their are at least 8 variants of *C. botulinum*) act by blocking neuromuscular transmission. The mechanism, which appears to be specific for peripheral neurons involves either blockage of the acetylcholine release mechanism at the nerve terminal or binding of the acetylcholine so that it cannot be released. Neuromuscular junctions that have been "poisoned" by the toxin can recover activity, a factor that has helped development of the use of "botox" treatment for many conditions (such as spasmodic torticollis) that result from hyperactivity of certain muscles. Injection of minute quantities of the botox into the selected muscles can alleviate the spasmodic condition and the reversibility guarantees that modest overtreatment will be "corrected" with time.

Items 504-508

The parents of a 6 year-old boy report that he has had repeated respiratory tract infections and persistent wheezing and coughing. The parents are of Swedish and German ethnic background and are healthy. The child shows elevated levels of salt in his sweat and thick respiratory mucous secretions. The child also shows pancreatic insufficiency due to obstruction of the pancreatic duct by mucus.

504.   The most likely diagnosis of this child's condition is

    (A)   cystic fibrosis
    (B)   duodenal atresia
    (C)   tetralogy of Fallot
    (D)   biliary atresia
    (E)   Tay-Sachs disease

505. This disease is caused by which genetic mechanism?

(A) autosomal dominant gene
(B) sex-linked recessive gene
(C) autosomal recessive gene
(D) translocation
(E) deletion

506. The protein primarily altered by different alleles of the affected gene in this disease is

(A) hexosaminidase A
(B) β-glucuronidase
(C) glucocerebrosidase
(D) α-amylase
(E) transmembrane conductance regulator

507. Cells from patients with this disease would show which response to elevated levels of adenosine 3′,5′-monophosphate?

(A) normal increase in Na^+ efflux
(B) lack of normal increase in Cl^- efflux
(C) down regulation of steroid receptors
(D) up regulation of steroid receptors
(E) increase in gluconeogenesis

508. Which plays the most direct role in protein phosphorylation **and** protein conformational changes?

(A) P_i
(B) AMP
(C) cAMP
(D) ADP
(E) ATP

ANSWERS AND TUTORIAL ON ITEMS 504-508

The answers are: **504-A; 505-C; 506-E; 507-B; 508-E**. This child suffers from **cystic fibrosis**, a common serious disease especially prevalent in Caucasians originating in Northern Europe. It is caused by an **autosomal recessive gene**. Many parents are asymptomatic heterozygous carriers. Approximately 1 in 2,500 newborns suffer from cystic fibrosis.

Typically, the respiratory and digestive tracts are most severely affected in cystic fibrosis. The airways of the respiratory system and ducts of exocrine glands in the digestive system are blocked by viscous mucus. This thick mucus impedes ciliary clearance of the respiratory system, leading to repeated infections (most commonly with *Pseudomonas*) and can compromise digestive function by limiting secretion of enzymes. Sweat glands are also affected in cystic fibrosis, leading to the secretion of sweat with increased salt concentration. This feature of the disease has been exploited for simple screening procedures.

Recently, the gene for cystic fibrosis has been identified and mapped to **chromosome 7**. The CF gene is large, containing about 250,000 base pairs, and encodes a protein of 1480 amino acids called the **cystic fibrosis transmembrane conductance regulator** (CFTR), a member of a large family of proteins involved in active transport of molecules across cell membranes. CFTR has two hydrophobic domains, each with six loops that span the cell membrane. It also has two nucleotide binding folds that bind and cleave ATP to release the energy required for active ion transport. This gene product also has a serine rich regulatory domain (R), thought to be the target of protein kinase activity. The most recent theory on how CFTR works involves several steps. When the R component is not phosphorylated, the Cl^- channel in the transmembrane domain of the molecule is closed. Cyclic AMP-dependent protein kinase activity leads to the phosphorylation of the R domain at the expense of ATP conversion to ADP. After phosphorylation of the R domain, the nucleotide binding folds are prepared to bind ATP. Binding and subsequent hydrolysis of ATP then causes a conformational change, leading to the opening of the Cl^- channel and a release of ADP and P_i. After release of bound nucleotide, the channel decays back to a closed state. Defects in the CFTR lead to decreased Cl^- efflux from cells and a generalized dehydration of mucous secretions in the body, especially in the respiratory and gastrointestinal systems. Dehydrated mucus has a higher viscosity and is thus more difficult to clear from the lungs, leading to blockage of airways and repeated respiratory infections.

Items 509-511

A 58 year-old alcoholic female shows hemorrhages near the base of hair follicles and swollen gingivae around her teeth. She has sores on her body that have healed poorly. A fractured toe has not healed properly after 3 months. Histological examination of sores reveals extensive granulation tissue with few collagen fibers. The diagnosis of scurvy is made.

509. The most altered molecule in scurvy is

   (A) fibronectin
   (B) elastin
   (C) collagen
   (D) cartilage proteoglycan
   (E) laminin

510. This condition is a the result of a dietary deficiency in

    (A)   vitamin A
    (B)   vitamin B_6
    (C)   vitamin C
    (D)   vitamin D
    (E)   vitamin E

511. The most accurate description of the fundamental lesion involved in the etiology of this dietary deficiency is

    (A)   deficient synthesis of collagen α chains
    (B)   leakage of capillaries
    (C)   decreased cross-linkage at desmosyl residues
    (D)   increased hydrolysis of tropocollagen telopeptides
    (E)   increased collagen turnover

ANSWERS AND TUTORIAL ON ITEMS 509-511

The answers are: **509-C; 510-C; 511-E. Vitamin C deficiency** results in a disease called **scurvy**. This disease is especially common in alcoholics because of dietary deficiency. Rupture of capillaries leads to follicular and gingival hemorrhages. Wound healing and fracture repair are also unusually slow in scorbutic patients. The granulation tissue at poorly healed wounds would have large numbers of fibroblasts, a cell type primarily involved in collagen biosynthesis.

Vitamin C is an essential cofactor in enzymatic reactions catalyzing the hydroxylation of prolyl and lysyl residues of collagen. Procollagen molecules without hydroxyproline residues have significant instability in their triple helices and are therefore more susceptible to degradation. In addition, extracellular collagen molecules will have few hydroxylysine residues and will thus be poorly cross-linked and therefore will be more susceptible to turnover. Also, fibroblasts secrete collagen more slowly in the vitamin C deficient state. Bone growth and fracture repair are also abnormal in scurvy.

Choose the best answer.

512. The HIV-1 virus causes AIDS. Which virus specific enzyme is most crucial for the replication of the HIV-1 genome?

   (A)   reverse transcriptase
   (B)   DNA polymerase
   (C)   RNA polymerase I
   (D)   RNA polymerase II
   (E)   DNA ligase

513. This unique enzyme has which peculiar function?

   (A)   has RNA cleavage activity
   (B)   utilizes ribonucleotide triphosphates
   (C)   makes specific cuts in DNA
   (D)   copies viral RNA into DNA
   (E)   joins the ends of viral RNA

514. This enzyme is targeted in antiviral therapy by which compound?

   (A)   5-FU
   (B)   methotrexate
   (C)   AZT
   (D)   α-amanitin
   (E)   rifampicin

ANSWERS AND TUTORIAL ON ITEMS 512-514

The answers are: **512-A; 513-D; 514-C**. After being internalized into the target cell, the **HIV** is uncoated in the host cell cytoplasm. The viral core particle includes the viral RNA genome which serves as a template for the synthesis of a single-stranded DNA complement that in turn functions as a template for a complementary DNA chain. The resulting double-stranded DNA, the so called proviral DNA, after insertion in a host cell chromosome, acts as a template for the synthesis of viral RNA and viral proteins utilizing the host-cell transcription and translation system.

The flow of genetic information, back from the viral RNA genome to the proviral DNA, is catalyzed by the virus-specific enzyme **reverse transcriptase**. This enzyme is named because its RNA-dependent DNA polymerase activity initiates the flow of information from RNA to DNA, opposite to the normal DNA to RNA information flow of the host cell. Since the synthesis of proviral DNA and its integration into the target cell chromosome is critical for viral expression and pathogenesis, the reverse transcriptase stage is an important target for antiviral therapy using **AZT** and ddI. Furthermore, reverse transcriptase activity, being largely a uniquely viral enzyme, is a useful marker for the late stage of viral infection in an infected individual.

Items 515-519

A twelve year-old boy is seen by his pediatrician with numerous freckles and depigmented areas on his face and hands. Closer examination reveals numerous basal cell carcinomas on his face. This child also has a long history of photophobia and currently has severe conjunctivitis.

515.   The most likely diagnosis of this disease is

    (A)   malignant melanoma
    (B)   vitamin D deficiency
    (C)   xeroderma pigmentosum
    (D)   shingles
    (E)   psoriasis

516.   The basic cellular mechanism **DEFECTIVE** in this disease is

    (A)   excision repair of damaged DNA
    (B)   activation of oncogenes by chemical carcinogens
    (C)   collagen cross-linking
    (D)   viral infection of epidermis, causing cell death
    (E)   RNA transcription

517.   Ultraviolet light induces which change in DNA?

    (A)   excision of pyrimidines
    (B)   formation of thymine dimers
    (C)   substitution of purines for pyrimidines
    (D)   frame shift mutations
    (E)   nonsense mutations

518. The mutation causing this disease is

   (A)   in enzymes for excision repair
   (B)   autosomal dominant
   (C)   sex-linked recessive
   (D)   sex-linked dominant
   (E)   pleiotropic

519. All of the following enzymes must function properly to reverse the genetic damage observed in this disease **EXCEPT**:

   (A)   DNA ligase
   (B)   RNA polymerase
   (C)   DNA polymerase
   (D)   endonuclease
   (E)   5'-exonuclease

ANSWERS AND TUTORIAL ON ITEMS 515-519

The answers are: **515-C; 516-A; 517-B; 518-A; 519-B.** This patient suffers from **xeroderma pigmentosum.** This rare malady occurs in approximately 1/250,000 people worldwide. It is caused by an **autosomal recessive mutation** found in all races of the human population. It is primarily manifested as an extreme sensitivity of the skin and other exposed portions of the body to ultraviolet light. In some varieties of xeroderma pigmentosum, systemic effects, e.g., damage to the central nervous system, are also observed.

   One mechanism for ultraviolet light-induced damage to DNA is the formation of thymine dimers between adjacent thymine residues of one strand of the DNA double helix. Several mechanisms for repair of damage to DNA have been discovered in bacterial systems. For example, excision repair utilizes a endonuclease which recognizes abnormal DNA near the 5' side of the thymine dimer, causing the damaged strand to separate from the double helix. Next, a DNA polymerase forms a new polynucleotide to replace the damaged sequence. Finally, a 5'-exonuclease removes the damaged sequence and a DNA ligase closes the break, resulting in the formation of a repaired duplex. In some cases of xeroderma pigmentosum, a genetic defect in **excision repair of damaged DNA** has been reported.

Items 520-524

At birth baby **X** was observed to have a toughened and rather inelastic superficial layer of the skin. This layer developed numerous small cracks and fissures. Twelve days later, the layer was "shed" and replaced by an apparently roughened and reddish outer layer that developed large scales.

520.   This child is most likely afflicted with

    (A)   bacterial meningitis
    (B)   lamellar ichthyosis
    (C)   Down syndrome
    (D)   idiopathic sepsis
    (E)   aplastic anemia

521.   Which of the following cells is most directly affected in this disease?

    (A)   fibroblast
    (B)   melanocyte
    (C)   hepatocyte
    (D)   mast cell
    (E)   keratinocyte

522.   At the level of specific enzymatic activities, the condition appears most closely related to the reduced activity of

    (A)   myosin ATPase
    (B)   ribosomal peptidyl synthetase
    (C)   transglutaminase
    (D)   tyrosine hydroxylase
    (E)   signal peptidase

523.   Select the organelle whose proper arrangement in the cell is most affected in this disease.

    (A)   centriole
    (B)   ribosome
    (C)   microtubule
    (D)   Golgi apparatus
    (E)   intermediate filament

524. In the normal course of events the organelle in Item 523 becomes cross-linked to which of the following?

(A) basal body
(B) mitochondrial outer membrane
(C) plasma membrane
(D) gap junction
(E) endoplasmic reticulum

ANSWERS AND TUTORIAL ON ITEMS 520-524

The answers are: **520-B; 521-E; 522-C; 523-E; 524-C**. The child is a victim of **lamellar ichthyosis**, a congenital disorder of the skin. The principal defect appears to be in the process by which keratinocytes become converted to heavily keratinized cells of the upper layers of the epidermis. Recent studies have shown that patients in several families with multiple occurences of the disease have altered or missing **transglutaminase**. This enzyme is required for proper binding of the **keratin intermediate filaments** of the stratum corneum of the epidermis to the surfaces of dying keratinocytes. Improper arrangement of these filaments leads to excess accumulation of incompletely cornified cells (failure of desquamation). This is turn results in thickened skin with large patches of excess stratum corneum and cracks between these plates. The cracks are sites of bacterial infection and/or water loss. Under normal conditions, transglutaminase is involved in the elaboration of a plasma membrane linked coating called the cell envelope, to which keratin intermediate filaments become bound. In the absence of proper cell envelope formation, keratin filaments do not become attached to the inner surface of the plasma membrane and this leads to incomplete shedding of dead surface cells. The progressive conversion of rapidly proliferating keratinocytes into highly modified cornified and dead squames that form a barrier to water flow and to bacterial invasion is a complex process of **programmed cell death**. A proper balance must be maintained between the function of the dead cells as parts of the barrier and shedding of the cells to make room for newly dying cells from more basal layers.

INDEX OF CLINICAL SCENARIOS AND DISEASES